MERCURY

MERCURY

☿

The Elusive Planet

Robert G. Strom

Smithsonian Institution Press
Washington, D.C.
London

Library of Congress Cataloging-in-Publication Data

Strom, Robert G.
Mercury: the elusive planet.
(Smithsonian library of the solar system)
Bibliography: p.
Includes index.
1. Mercury (Planet). 2. Project Mariner.
I. Title. II. Series.
QB611.S77 1987 523.4′1 86–26030
ISBN 0–87474–892–5

British Library Cataloging-in-Publication Data is available.

The paper used in this publication meets the minimum requirements of the American National Standard for Permanence of Paper for Printed Library materials Z39.48–1984.

Cover: Photomosaic of Mercury viewed by Mariner 10 as it left the planet after the first encounter. The color is not real.

Frontispiece: Mercury's enormous core and thin lithosphere are shown in perspective with this photomosaic taken by Mariner 10. (Artwork by Karen Denomy.)

This book is dedicated to the memory of Gerard P. Kuiper, a member of the Mariner 10 Imaging Team and father of modern planetary science, who introduced me to the Solar System.

Contents

Foreword

During the past twenty-five years, the study of the planets and other bodies in the Solar System has expanded from the realm of astronomers into that of geologists, meteorologists, chemists, and physicists. These scientists have been able to gather a wealth of data by means of numerous manned and unmanned spacecraft launched during this period. Although these voyages provided the basis for futher exploration of the Solar System, in some cases they occurred so frequently that there was little time to assimilate all the information returned. With the hiatus in planetary missions during the mid-1980s, however, scientists have been able to concentrate their attention on analyzing the data collected thus far.

All the inner planets except Mercury have been explored by more than one spacecraft. Despite this seeming inattention to the innermost planet of the Solar System, the results of the three flybys of Mariner 10 provided a strong base of information for developing a detailed history of that planet's evolution, and for formulating future mission scenarios. As you will discover in this volume, several constrasting theories of Mercurian history exist, and it is these questions that will inspire our return to Mercury.

The next decade will see a rebirth of exploration of the planets and other bodies in the Solar System, as plans are now being finalized for the Galileo Mission to Jupiter, the Mars Observer, the Magellan Mission to Venus, and the Comet Rendezvous/Asteroid Flyby Mission. We hope that the "Smithsonian Library of the Solar System" will help a wide audience to understand why these missions are needed and how new discoveries will change our concepts of the evolution of our neighbors in space, and of the evolution of the Earth itself.

This series would not be possible without the work of the dedicated scientists and engineers who plan, develop, and carry out the exploration of the Solar System. We are indebted to the National Aeronautics and Space Administration for supporting these goals and the scientific research that provides the basis for future missions. We are also grateful to staff of the National Air and Space Museum and to Felix C. Lowe, director of the Smithsonian Institution Press, for their support of the series.

Ted A. Maxwell, Series Adviser
Center for Earth and Planetary Studies
National Air and Space Museum

Preface

Mercury: the swiftest planet, messenger to the Greek and Roman gods. This tiny planet has evoked the curiosity of both the ancients and modern-day scientists. Before the flight of Mariner 10 in 1973–75, very little was known about the Sun's nearest neighbor. The difficulty of observing Mercury from Earth frustrated astronomers, who were led to many erroneous conclusions.

Mariner 10 provided us with our first and, to date, only close-up look at Mercury. Although fraught with difficulties and near calamity, this mission increased our knowledge of the planet a thousandfold. Many unanswered questions remain that can only be answered by more sophisticated explorations, but currently there are no plans to send additional spacecraft to Mercury during the remainder of this century.

Superficially, Mercury looks like the Moon, and one may be led to believe that they share a similar history, but we now know this is not true. In many respects, Mercury is quite different from the Moon. Some of these differences are unique in the Solar System and provide important insights into the evolution of the terrestrial planets. This neglected child of the Solar System, with its pockmarked and wrinkled face, has a fascinating story to tell.

This book aims to point out the uniqueness of Mercury and to pass on some of the excitement I have experienced while studying this unusual planet. It will lead you to a curious shrunken world with an iron core larger than our Moon; a world that once bubbled with volcanic activity, and where huge planetesimals smashed into the surface; a world where gigantic cliffs up to three kilometers high cut the surface for hundreds of kilometers; a world where the temperature climbs to the melting point of zinc and then plunges to the freezing point of methane (natural gas); a world whose day is two years long.

Without the help of many people, this book would not have been possible. I sincerely thank Bruce Hapke of the University of Pittsburgh and Paul Spudis of the U.S. Geological Survey, Flagstaff, for reviewing the manuscript and making many helpful suggestions. Jim Dunne at the Jet Propulsion Laboratory reviewed the chapter on the flight of Mariner 10 and provided many of the illustrations for that chapter. I also thank Eugene Levy and Charles Sonett for reviewing the chapter on Mercury's interior and magnetic field. Ewen Whitaker helped with several illustrations and provided many useful suggestions for improving the manuscript. Karen Denomy of the Lunar and Planetary Laboratory, University of Arizona, prepared the artwork for the frontispiece. Ted Maxwell at the National Air and Space Museum was instrumental in guiding the book through the publication process and provided many helpful suggestions. Finally, I sincerely thank Mary Ellen Byers for typing the manuscript and offering many useful suggestions for its improvement.

Chapter 1

The Elusive Planet

Huddled around smoldering fires just before dawn, prehistoric men and women must have noticed an exceptionally bright "star" that climbed above the amber glow of the rising Sun until it faded and disappeared in the brilliance of a new day. At other times of the year, when they returned from the hunt and watched the Sun dip beneath the horizon, a similar "star" would appear and grow brighter as it followed the Sun's path, lingering in the night sky and finally disappearing beyond the horizon. Later civilizations would call these morning and evening objects the planet Venus.

During certain times of the year, especially spring and autumn, primitive people probably noticed another bright, but fainter, object that quickly followed the Sun over the horizon at sunset, or disappeared in the bright light of dawn shortly before sunrise. We would later call this object Mercury. They might have noticed that over time both of these objects moved with respect to the fixed stars and therefore differed from them; other "stars" may have been observed to wander among the fixed stars, but only these two would be associated with either the rising or setting Sun. Early peoples therefore may have seen some special, ritual significance in these "stars." Many thousands of years later, their descendants would build a ship that would sail through space to visit both of these celestial bodies. This vessel—the Mariner 10 spacecraft—would reveal a world that has become more familiar to us than the Earth was to the ancients.

Early Pre-Telescopic Observations

Mercury played an important part in the religious lives of early civilizations. The ancient Egyptians called it Sobkou and were the first to discover that, like Venus, it travels around the Sun. Mercury was known as Bi-ib-bon to the Sumerians, while the Assyrians and Chaldeans called it Goudond. The Chaldeans represented Mercury as wearing a royal robe and tiara, and often with four wings indicating rapid motion. To them Mercury was good when among benevolent stars and bad when among malevolent ones. The Babylonians had several names for Mercury, one of which was Nabou, the ruler of the universe, who alone was able to raise the Sun from its bed. In Phoenicia, Mercury was called Mokim or Monim, and was either good or bad depending on circumstances. The ancient Scandinavians named it Odin, while early Teutonic peoples called it Woden; both stood for the father of the gods.

The ancient civilizations of the Middle East knew that the time between the reappearance of Mercury in the same configuration in the sky was shorter than for the other planets, and correctly reasoned that it therefore moved more rapidly. This time interval of 115 days, called the synodic period, was accurately measured by the Greek astronomer Eudoxus in about 400 B.C. As seen from Earth, when Mercury reaches its largest angular distance from the Sun it is said to be at greatest elongation. There was confusion among the early Greeks concerning the eastern elongation of Mercury when it is an evening "star" and the western elongation when it is a morning "star," which led them to believe these "stars" were two distinct objects. During the time of Plato, however, in about 350 B.C., the Greeks recognized that the two "stars" were in fact one planet—something the ancient Middle Eastern civilizations had known much earlier.

The Greeks observed that differences between Mercury's greatest elongations, and also its angular deviations from the ecliptic (the Sun's path through the sky), were greater than those of any other planet. Because this planet trav-

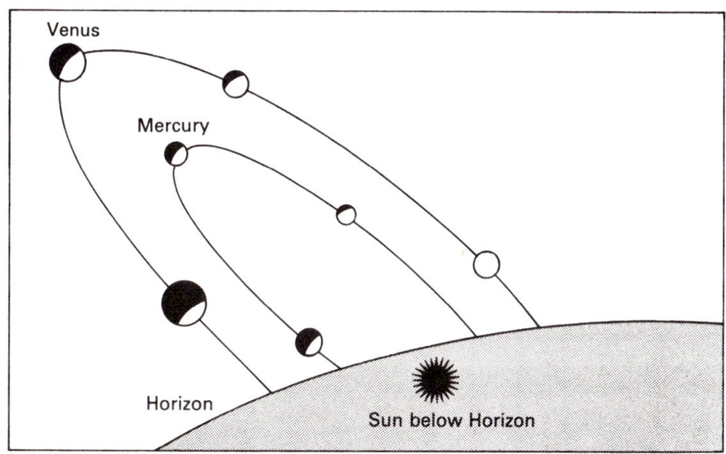

As seen from Earth, both Mercury and Venus stay close to the Sun because they orbit the Sun within Earth's orbit. At their largest angular distance from the Sun as seen from Earth, they are at greatest elongation. In this diagram, the two planets are shown at eastern elongation when they set after the Sun and appear as evening "stars."

eled more rapidly than the others, the Greeks named it Hermes, the messenger of the gods and the god of twilight and dawn who announced the rising of Zeus, the god of day. (The adjective Hermian is used when referring to certain aspects of Mercury, such as cartographic designations.) The Romans largely adopted the Greek gods as their own, but changed their names. Our modern name for Mercury is derived directly from the Latin name Mercurius, the Roman designation for Hermes. Its astronomical symbol ☿ can be traced to a medieval Greek manuscript, where it takes the form ☿. The horizontal cross is a modern addition. The "horns" at the top of the symbol represent the speedy Mercury's wings.

The English word for the midweek day of Wednesday also derives from the planet Mercury. English words for the days of the week are based on old Teutonic equivalents of the Roman names for the Sun, Moon, and planets. These Germanic peoples called the midweek day "Wodenes Daeg," or Woden's day—their name for Mercury.

The chemical element Mercury, commonly called quicksilver, is so named because its rapid flow suggested the fleet-footed planet. Mercury's chemical symbol Hg derives from the Greek name for the substance, Hydrargyrum or "water silver."

Telescopic Observations

In the early 1600s, the Copernican or heliocentric view of the Sun at the center of the Solar System became established through the observations and calculations of Tycho Brahe, Galileo, and Johannes Kepler. It was then realized that Mercury was the innermost planet, and that the large difference between its greatest angular distances from the Sun and its angular deviations from the Sun's path in the sky, observed by earlier astronomers, were due to an extremely elliptical orbit and a large orbital inclination with respect to Earth's plane of motion around the Sun. Not until the discovery of Pluto, nearly 300 years later, would a planet be found to have a more elliptical or highly inclined orbit.

The invention of the telescope ushered in the modern era of astronomy. The telescope was first used by the English scientist Thomas Harriot in August 1609 and later that same year by the Italian astronomer Galileo. Mercury, however, proved to be an elusive object for telescopic observations because it is never more than 28 degrees from the Sun as viewed from Earth. Thus, observations must be made during daytime or at twilight through a long path-length of the Earth's atmosphere. This results in extremely poor "seeing" conditions that severely degrade the ability to resolve surface features. There are also only about thirty or forty days during the year when Mercury can be seen as either a morning or an evening "star," which severely limits the amount of time it can be observed. Even our modern-day observations with large telescopes

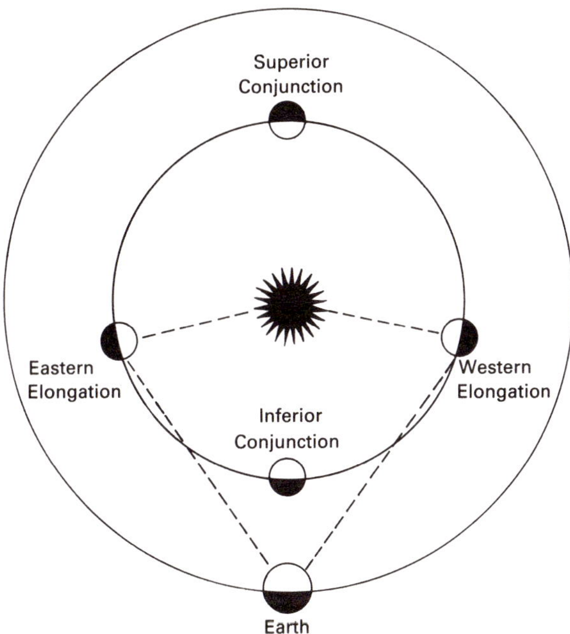

Venus and Mercury are called inferior planets because their orbits lie between the Earth and Sun. When Mercury or Venus is between the Earth and Sun, it is said to be at inferior conjunction. When on the far side of the Sun they are at superior conjuction. Sometimes at inferior conjunction the nodes of the orbital planes align, and Mercury or Venus cross or transit the Sun's disk to appear as dark dots.

yield only scanty information about Mercury. For example, its true rotation period was not known until 1965, and this information was obtained not by telescopes, but rather by radar observations. Nevertheless, telescopic observations did yield information that indicated Mercury was an unusual planet deserving closer study.

Venus and Mercury are the only planets to show phases similar to the Moon's, because they have orbits that lie within Earth's orbit—they are "inferior" planets. One of the principal arguments against the Copernican Theory was that no such phenomenon had been observed for either planet. But naked-eye observations could never de-

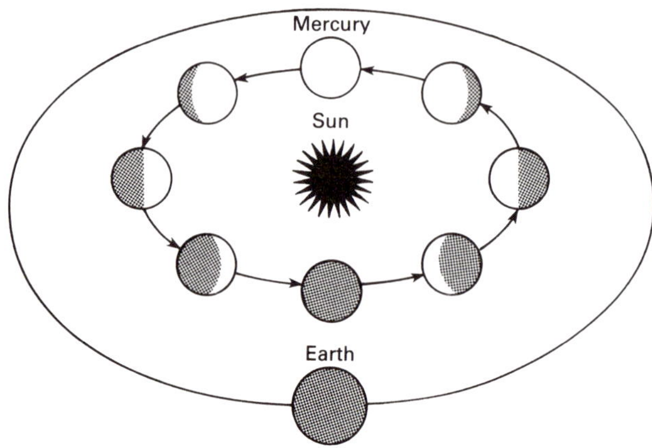

In 1639 the Italian astonomer Giovanni Zupus discovered that Mercury shows phases like those of the Moon and Venus.

tect phases on these objects. When Galileo turned his newly invented telescope to Venus in 1610, he discovered that that planet has phases, but he was unable to detect phases on Mercury because his telescope was not powerful enough. In 1639 an Italian named Giovanni Zupus, using a more powerful telescope, discovered that Mercury also has phases. At the time, these observations were of profound significance because they provided strong evidence that the Copernican Theory was correct; indeed, the planets, including the Earth, did circle the Sun. Our conception of the universe changed dramatically; we were no longer at its center.

Since Mercury, like Venus, is an inferior planet, it should move across or transit the face of the Sun as seen from Earth. Based on his newly devised laws of planetary motion, Kepler in 1627 was the first to predict the time when Mercury would transit the Sun. This prediction enabled the Frenchman Pierre Gassendi to first observe a transit on November 7, 1631. In 1661 the Polish astronomer Johannes Helvelius observed another transit; Mercury even received royal attention when King Louis XV of France observed the transit of May 6, 1753, at Mendon Castle.

These three photographs record the progress of the planet Mercury across the disk of the sun during the transit of May 9, 1970. The first image shows Mercury superimposed over a large sunspot. The next two images show the movement of the planet over a period of one hour. A transit can occur only during inferior conjunction, when Mercury is at or near a node in its orbit. (Courtesy Geoffrey Chester, National Air and Space Museum.)

Transits of the Sun by Mercury are rather rare events. Because of its short orbital period, Mercury often passes between the Earth and Sun (called inferior conjunction). The planet, however, usually passes above or below the Sun as seen from Earth because of the high inclination (7 degrees) of its orbit. The points where Mercury's orbit cross the plane of the ecliptic are called the nodes of its orbit. A transit will occur only when Mercury happens to be quite near a node during inferior conjunction. As a consequence, the minimum time between successive transits is three years, and the maximum is thirteen years. Therefore, only about thirteen transits of Mercury occur each century. The next transit will occur on November 6, 1993, but it will be visible only in Asia, Australia, and the Indian Ocean. The next transit visible in North America will not take place until November 15, 1999.

Mercury's small size was not realized until the mid-1700s, when telescopes became large enough to make relatively accurate measurements. At this time the measured diameter ranged from about 4,400 to 6,500 kilometers,

The size of Mercury compared to the Moon and other terrestrial planets, the Galilean satellites of Jupiter, and the Titan satellite of Saturn. Mercury is about a third the size of Earth and almost excactly the same size as Callisto. (Courtesy NASA/Goddard Space Flight Center.)

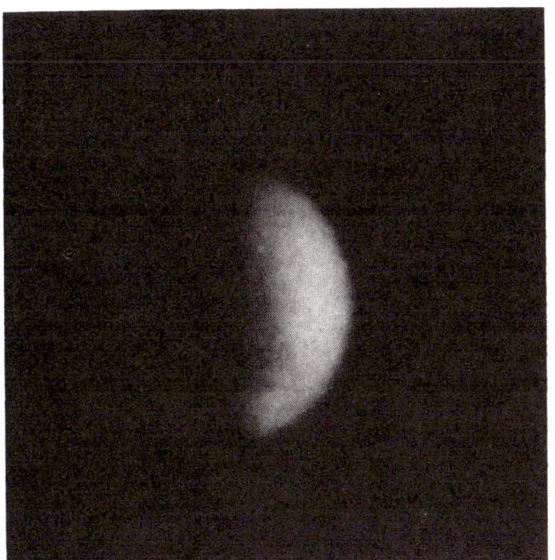
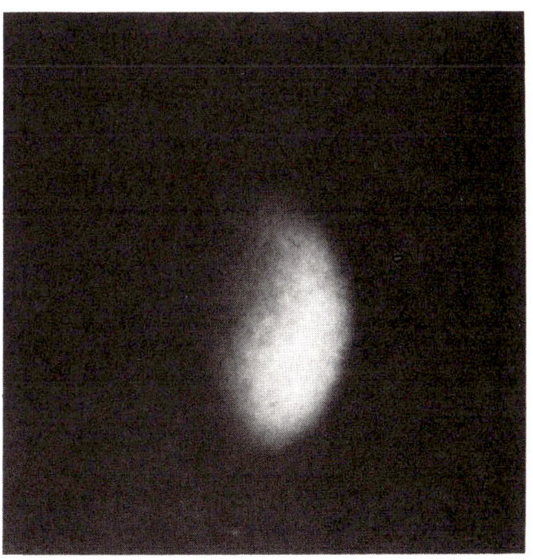

These Earth-based telescopic photographs of Mercury show several of its phases, but only indistinct surface markings because of its small size, great distance from Earth, and closeness to the Sun. (Courtesy New Mexico State University.)

which established Mercury as the smallest planet yet discovered. Modern telescopic measurements yielded a more accurate diameter of about 4,800 kilometers—close to the value of 4,878 kilometers determined by the Mariner 10 spacecraft in 1974. Mercury was considered the smallest planet in the Solar System until recent measurements (1980) showed Pluto to have a diameter between 3,000 and 3,600 kilometers, or about the size of our Moon.

Mercury's small size, together with the poor observing conditions from Earth, make it extremely difficult to detect surface markings. This elusive planet is only visible low in the sky during morning and evening twilight, and the poor observation conditions that prevail near the horizon rarely allow even the largest telescopes to show details on Mercury's tiny disk. When viewed during daylight, the brightness of Earth's sky usually exceeds that of Mercury, thereby reducing the contrast of its markings close to the threshold of visual perception. If the sky is hazy or the telescope optics slightly dusty or seeing conditions less

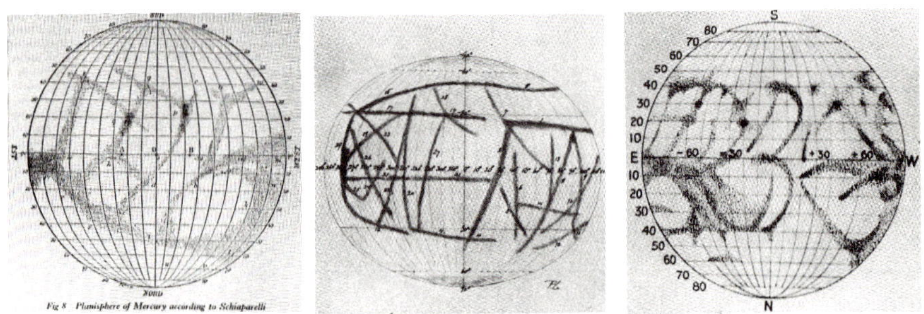

Left: Giovanni Schiaparelli's map of Mercury (1889). Middle: Percival Lowell's map of Mercury, drawn in 1896. Right: The René Jarry-Desloges map of Mercury (1920).

than excellent, Mercury always appears featureless. Even the best telescopic photographs taken under optimum conditions show only vague hints of markings. Is it any wonder that early astronomers were so frustrated in their attempts to observe Mercury?

The only visual method for determining a planet's rotation rate is to measure the change of position of its surface markings at different times. For Mercury, this method was doomed to failure because of the poor definition of its surface features and the peculiar relationship between its rotation period and synodic period.

The first observation of Mercury's surface features is attributed to the German astronomer Johann Hieronymus Schröter, who detected dusky markings in the late 1700s. Because the surface markings did not appear to change from day to day, Schröter, and in 1867 the astronomer Wilhelm Prinz, assumed that Mercury rotated with a period of twenty-four hours—the same as Earth. In Milan, Giovanni V. Schiaparelli, who had become famous for his observations of the so-called Martian canals, observed Mercury from 1882 to 1889. During this time, he was unable to detect any significant change in the position of a particular feature across the planet's disc. Based on the position of the features on the planet with respect to Mer-

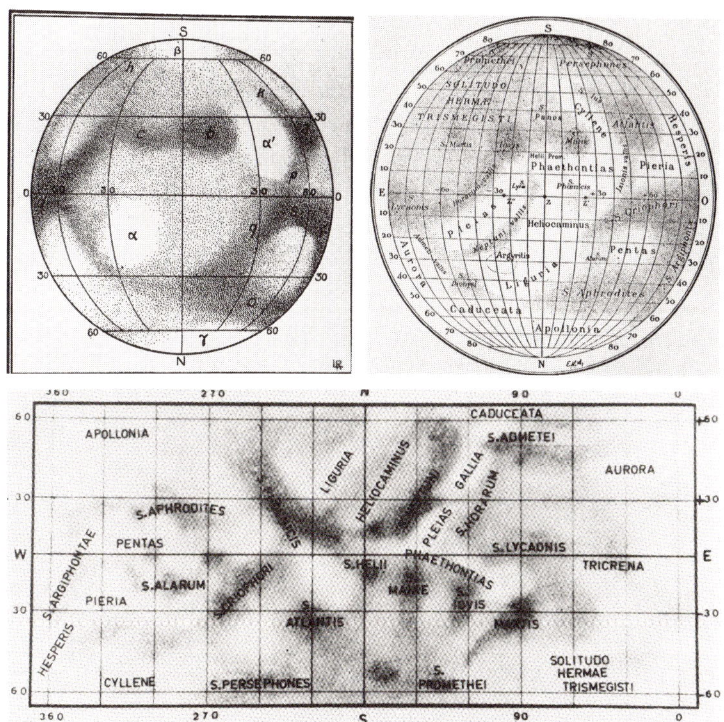

Top left: Map of Mercury by M. Lucien Rudaux (1927). Top right: Eugene Antoniadi's map of Mercury (1934). Bottom: A pre-Mariner 10 contemporary map of Mercury by Audouin Dollfus (1965).

cury's location in its orbit, Schiaparelli concluded that Mercury was in synchronous rotation. In other words, it was thought to turn once on its axis during the time it takes to complete one orbit around the Sun. Since Mercury's orbital period is eighty-eight days, then one rotation of Mercury would also take eighty-eight days. Thus, Mercury would continually keep the same side facing the Sun, just as the Moon always keeps the same face toward Earth. Unlike the Moon, however, one hemisphere of Mercury would experience perpetual day and the other hemisphere would remain in eternal darkness. Because the

Moon orbits the Earth, which, in turn, orbits the Sun, the entire lunar surface is illuminated during a single orbit around the Earth.

By the early 1960s about twenty maps of Mercury had been compiled, some based on extensive visual observations by experienced astronomers such as Eugene Antoniadi, Georges Fournier, and Audouin Dollfus. All appeared to be consistent with an eighty-eight-day rotation period. Then in 1965, Gordon Pettengill and Rolf Buchanan Dyce, using the Arecibo radar facility in Puerto Rico, discovered that Mercury's rotational velocity was consistent with the rotation period of about fifty-nine days. Subsequent radar observations confirmed this result and established that the planet has a rotation period of 58.6 days—not the synchronous period of eighty-eight days indicated by the visual and photographic observations.

What went wrong? Were the visual observations worthless, or was Mercury playing some cruel trick on us? It turns out that the visual observations are consistent with both an eighty-eight- and fifty-eight-day rotation period. A 58.6-day period is exactly two-thirds of the planet's eighty-eight-day year. This means that Mercury rotates three times on its axis every two orbits around the Sun. As a consequence, after three synodic periods we see the same face of the planet at the same phase. Because three synodic periods also corresponds with the time interval between favorable observing conditions—when most visual observations were made—an observer working only at such times will see the same surface features in the same places for six consecutive years. This was precisely the criterion used to justify an eighty-eight-day period! Eventually this periodicity gets out of phase, but because most maps are usually based on only a few years of observations, it is not surprising that most of an observer's drawings are consistent with one map.

A planet's mass is determined from analysis of its gravitational effect on the positions and orbits of other planets, asteroids, comets, or spacecraft. These gravitational effects are called perturbations, and, if the motions are well known, they can offer a powerful means of measuring planetary masses. The first experimental determination of Mercury's mass was made by Johann Franz Encke in 1841, when he measured the perturbation by Mercury of the comet that now bears his name. During the following forty years, other mass determinations were made principally by measuring the perturbations of Venus. All these values were within about 20 percent of the best modern telescopic measurements, which yielded a mass of 3.3×10^{26} grams. This value is quite close to the more accurate one determined by the Mariner 10 spacecraft.

Although the mass of Mercury is small compared to that of other planets, its mass is much greater in comparison to its size than that of most other planets and satellites. This fact indicates that Mercury contains a larger proportion of heavier elements than other planets and, therefore, is unique in the Solar System. As we shall see later, this property has had profound effects on the planet's history.

By studying the manner in which light is reflected from a surface, it is possible to learn something about the physical properties and composition of that surface. The brightness of Mercury varies depending on its distance from Earth. At its brightest, Mercury shines more brilliantly than Sirius, the brightest of all the fixed stars. Its brightness is surpassed only by Venus and Jupiter, and occasionally by Mars when it is close to Earth. Even at its faintest, Mercury is still as bright as the stars Deneb and Aldebaran. Knowing Mercury's brightness at a particular distance, it is possible to calculate its albedo—the percentage of light reflected from the surface. It turns out that Mercury is a relatively dark object that reflects only about 8 percent of the light that it receives from the Sun, a value somewhat greater than for our own Moon. The amount of light received by an observer from a planet depends on the phase angle, which is the angle between the Sun and the observer, as seen from the planet. For example, if the planet appears half lit, then the Sun-planet-observer angle, or phase angle, is 90 degrees; if it is fully lit, then the phase angle is zero degrees. For Mercury, the amount of light reflected at various phase angles is similar, but not identical, to that of the Moon. This pattern indicates that, like the Moon, Mercury is covered by a dusty fragmental layer termed a regolith.

During the nineteenth century, Mercury was thought to possess a dense atmosphere comparable to Earth's. This erroneous conclusion was largely based on poor spectroscopic evidence obtained by the German astronomer Hermann Karl Vogel in 1871. Later astronomers doubted Mercury had any atmosphere at all, because of its low albedo and the fact that the temperatures are so high on Mercury that gas molecules would be accelerated to speeds large enough to escape the planet. Although a dense atmosphere was ruled out, some astronomers as late as the 1950s and 1960s claimed that Mercury possessed a tenuous atmosphere comparable to that of Mars. This conclusion was based on supposed obscurations of surface features, and certain differences in the way light is polarized at Mercury. But these observations were marginal at best. Nevertheless, there was still enough evidence to suggest that Mercury might have a tenuous atmosphere. This question remained open until the flight of Mariner 10 in 1973.

Pre-Mariner 10 Speculations

The accumulated knowledge of Mercury from our best Earth-based observations remained woefully small. In many ways, we knew little more about Mercury than did Copernicus. The observations did show, however, that Mercury was unique in the Solar System in some respects and similar to the Moon in others: Mercury's orbit was more elliptical and inclined than any other planet except Pluto (which may, in fact, be an escaped satellite of Neptune). Unlike other planets, Mercury's rotation period was locked into a 3:2 resonance with its orbital period. Its mass was larger in proportion to its size than any other planet or satellite. This implied an unusual internal constitution probably consisting of an extraordinarily large iron core compared to its diameter. On the other hand, Mercury's size was somewhat comparable to the Moon's, and it reflected light in a similar manner.

These facts raised numerous questions about Mercury that could not be answered by further Earth-based observations. By late 1969, we had detailed knowledge of the Moon from the unmanned Lunar Orbiter and manned Apollo missions and our first close-up look at Mars from the Mariner 4, 6, and 7 missions. Would the surface of Mercury look like the Moon or Mars, or would it be quite different from anything we had ever seen? What influence would a large iron core have on geologic processes? Did Mercury have a magnetic field, like Earth, or essentially none, like the Moon? Would there be a tenuous atmosphere on Mercury, and if so, what would be its composition and how would it be maintained? How would Mercury's history compare with those of other terrestrial planets?

Speculations on these subjects were rampant. Some believed that Mercury's surface would be as smooth as a billiard ball due to an early, intense flare-up of the Sun called a T-Tauri phase. According to this idea, because Mercury is so close to the Sun its outer layers would have been stripped away by intense solar activity, leaving behind a relatively thin crust and a featureless surface. This explanation could account for the proportionately large core and thin crust. Others speculated that the surface would be cratered like the Moon and Mars, but that there would be far fewer craters. They assumed that the craters on the Moon and Mars were created by impacts of asteroids, and since Mercury was so far from the asteroid belt there should be fewer impacts and therefore fewer craters. Some believed that Mercury's thin crust and large core would produce conditions that would cause crustal

The small dot in the sky is Mercury, photographed by Surveyor 7 from the surface of the Moon on January 23, 1968. The fuzzy glow at the horizon is the Sun's corona. Mercury's elongated shape is due to its motion during this thirty-minute exposure. (Courtesy NASA.)

movements similar to those on Earth. Accordingly, there might be large mountain ranges and the Mercurian equivalent of mid-ocean ridges caused by Earth-like plate tectonics. At that time, it was believed that planetary magnetic fields could be generated only if the planet rotated relatively fast and if its core was largely molten and consisted of a metallic substance. Mercury, however, rotates slowly, and thermal history models suggested that the core would have solidified long ago. This factor prompted most scientists to believe Mercury would not have an appreciable magnetic field and that the solar wind would interact with Mercury in the same way that it does with the Moon. Almost all these speculations would prove to be wrong.

The only way of providing answers to these fundamental questions was to fly a spacecraft to Mercury equipped with an array of instruments capable of obtaining the necessary data. Up to this time, only one spacecraft had taken a picture of Mercury. On January 10, 1968, the Surveyor 7 spacecraft landed on the Moon near the crater Tycho to take pictures of the surface and make chemical and mechanical measurements of the soil. During this mission, the spacecraft photographed the Sun's corona on January 23, and in these pictures Mercury appears in the sky. These images, however, were hardly sufficient to provide any meaningful information about the planet. Thus, NASA initiated the Venus/Mercury Project in 1970, and work began on building the Mariner 10 spacecraft to fly this mission. Although fraught with difficulties and near disaster, Mariner 10 would return an abundance of information about Mercury, making this poorly understood planet a more familiar world, rich with scientific information disproportionate to its size.

Chapter 2

Being There: Flight to Mercury

In 1962 it was discovered that the positions of the Earth, Venus, and Mercury would be ideally configured in 1970 and 1973 so that a spacecraft launched to Venus could be perturbed by the planet's gravity field onto a new trajectory that would take it to Mercury. Thus, with a single spacecraft it would be possible to visit two planets with a minimum expenditure of on-board fuel. It was crucial that the launch take place in either 1970 or 1973, because a similar economical opportunity would not occur until the mid-1980s. In 1968 the Space Science Board of the National Academy of Sciences endorsed a mission to fly by Venus and Mercury at the 1973 opportunity. Late the next year, Congress approved the mission to begin development in 1970.

Concepts of a Space Mission

The National Aeronautics and Space Administration (NASA) designated the Jet Propulsion Laboratory (JPL) of the California Institute of Technology to develop and operate the mission. A Mariner-type spacecraft was chosen for the mission, and the project was named the Mariner Venus/Mercury Mission. The spacecraft would later be called Mariner 10 because it was the tenth Mariner vehicle to be launched.

Although this spacecraft had to penetrate and operate in a more hostile environment than any previous spacecraft, NASA insisted that it initiate a new breed of low-cost mis-

"Three ships is a lot of ships. Why can't you prove the world is round with one ship?"

To hold down the cost of the Venus/Mercury mission, NASA insisted that only one spacecraft be used to explore both planets. (Courtesy Jet Propulsion Laboratory.)

sions. One of the most severe restrictions was the use of only one spacecraft to explore both planets. Previous lunar and planetary missions had used two or more spacecraft to gather more data and provide a backup in case one failed. The Mariner 8 spacecraft to Mars, for instance, had experienced a launch failure, leaving only Mariner 9 to carry out the first orbital mission to that planet. Fears of a similar failure prompted the Mariner Venus/Mercury Project to request a backup spacecraft when it became evident that one could be prepared within the total project cost of $98 million. Basically, a backup involved only a small increase in the number of spare components. However, this backup spacecraft was to be used only if a failure of the prime spacecraft or launch vehicle occurred during the October 16 to November 21 launch window. If the primary spacecraft failed on or after November 21, the

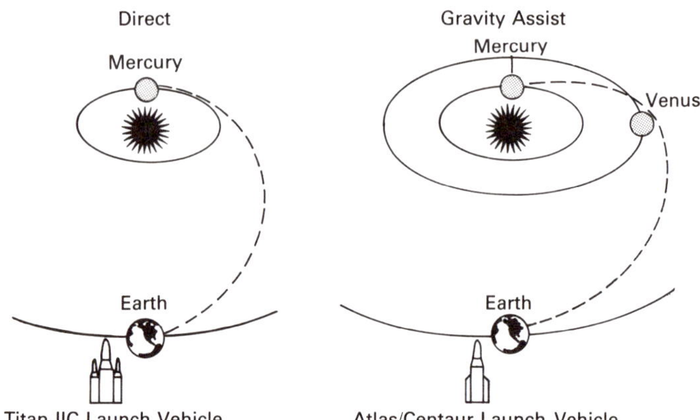

The gravity-assist trajectory to Mercury uses the gravity and orbital motion of Venus to propel a spacecraft into the inner Solar System without the need to expend precious fuel except for minor trajectory corrections. A direct flight without a Venus assist would require a much larger launch rocket to deliver the same payload of instruments. Mariner 10 was the first planetary mission to use the gravity-assist trajectory, which has since been used by Voyager at Jupiter, Saturn, and Uranus for its exploration of the outer Solar System.

second spacecraft could not be launched and the mission would be lost. Understandably, the project scientists and engineers were extremely concerned.

The gravity-assist trajectory technique was needed to obtain an economically acceptable mission. This technique allows a spacecraft to change both its direction and speed without using valuable fuel, thereby saving time and increasing the size of the scientific payload. With its use, the Mariner spacecraft could be launched with an acceptable payload by an Atlas/Centaur rocket. For a direct flight to Mercury, a much larger and more costly Titan III C/Centaur would be required, and its high speed past Mercury would permit only a short time to acquire data. The Venus/Mercury gravity-assist trajectory enabled one spacecraft to explore two planets, and it also provided a bonus return visit to Mercury.

At a conference on the Venus/Mercury mission held at JPL in Pasadena, California, the late Giuseppe Colombo of the Institute of Applied Mechanics in Padua, Italy, noted that after Mariner 10 flew by Mercury, its orbital period around the Sun would be quite close to twice Mercury's orbital period. He suggested that a second encounter with Mercury could therefore be accomplished. A detailed study of the trajectory by JPL confirmed Colombo's suggestion and showed that by carefully selecting the Mercury flyby point, a gravity turn could be made that would return the spacecraft to Mercury six months later. In fact it would be possible to achieve multiple encounters with Mercury; the number would depend on the fuel available for midcourse corrections and attitude control. Mariner 10 would eventually achieve three encounters with Mercury before its fuel ran out.

The Flight Plan

The mission plan for Mariner 10 was the most complex for any planetary mission flown up to that time. It called for a launch sometime between October 16 and November 21, 1973. November 3 was eventually chosen, so that the spacecraft would encounter Mercury at a time when it could view the planet about half lit (called quadrature). Viewing Mercury at this phase would make it easier to distinguish surface features by their shadows. The trajectory relied on Venus's gravitational field to alter the spacecraft's flight path and speed relative to the Sun. Properly aimed, the spacecraft's speed would be reduced, causing it to fall closer to the Sun and cross Mercury's orbit at the precise time needed to encounter the planet.

New levels of accuracy were required to intercept Venus with high precision. The flyby point at Venus had to be controlled within 400 kilometers, or a Mercury encounter

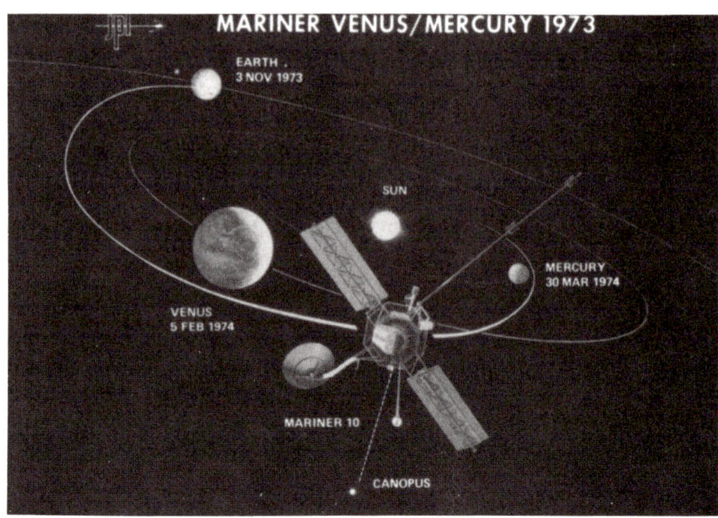

This illustration shows the launch from Earth and the flybys of Venus and Mercury by Mariner 10. The times of launch and arrival at the planets were precisely defined. (Courtesy Jet Propulsion Laboratory.)

would not take place. At least two maneuvers would be needed between Earth and Venus, and probably two more between Venus and Mercury. Mariner 10 would be the first planetary mission to use this gravity-assist technique.

The flight plan called for the upper-stage Centaur rocket to be turned off for twenty-five minutes shortly after launch from the Kennedy Space Center. This would place it in a parking orbit that would carry it partway around the Earth. Then a second ignition would thrust the Mariner spacecraft in a direction opposite to the Earth's orbital motion, providing the spacecraft with a lower velocity relative to the Sun than the Earth's orbital velocity and allowing it to be drawn inward by the Sun's gravitational field to achieve an encounter with Venus. After a few months Mariner 10 would approach Venus from its night side, pass over the sunlit side, and, slowed by Venus's gravitational field, fall inward toward the Sun to rendezvous with Mercury.

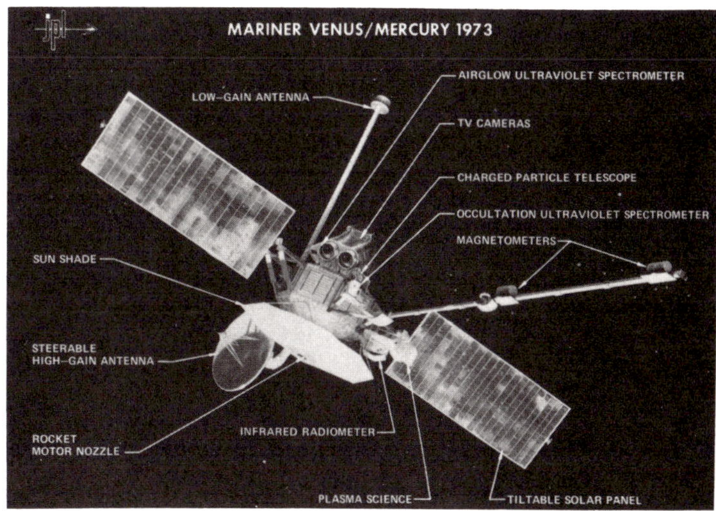

This illustration of Mariner 10 shows the two solar panels, the long magnetometer boom protruding to the right, the sun shield and high-gain antenna at the bottom, and the twin Cassegrain telescopes that provided the high-resolution pictures of Venus and Mercury. Other instruments are attached to the body of the spacecraft. The long, polelike object at the top is the omnidirectional, low-gain antenna.

The Spacecraft

The Mariner 10 spacecraft evolved from more than a decade of Mariner technology, begun with the Mariner Venus mission in 1962 and culminating with the Mariner Mars Orbiter in 1971. Like other Mariners, it consisted of an octagonal main structure, solar cells and a battery for electrical power, nitrogen gas jets for three-axis attitude stabilization and control, star and Sun sensors for celestial reference, an S-band radio for command and telemetry, a high- and low-gain antenna, a scan platform for pointing scientific instruments, and a hydrazine rocket propulsion system for trajectory corrections.

Mariner 10, however, would fly much closer to the Sun than any previous spacecraft. It would be subjected to solar intensities up to four-and-a-half times greater than at Earth, requiring thermal control to maintain temperatures at a level that would not damage the spacecraft systems. To meet this requirement, a large sunshade, louvers and protective thermal blankets, and the ability to rotate the solar panels were added. Because Mariner 10 would be so close to the Sun, only two solar panels were needed to generate electricity to power the spacecraft. As the spacecraft approached the Sun, the panels could be rotated to change the angle at which light fell on them, and thus maintain a suitable temperature—about 115° C.

Another major design change from past Mariners was the addition of a capability to handle up to 118,000 bits per second of television data. If this high data rate had not been implemented, all television pictures of Mercury would necessarily have been recorded on tape and played back at later times, as occurred on previous planetary missions. Since the tape recorder was capable of holding only thirty-six pictures, the amount of high-resolution coverage of Mercury taken during the short interval near closest approach would have been severely limited. The high data rate permitted the spacecraft to transmit high-quality pictures in "real time," or as fast as they were taken, thereby allowing hundreds of images to be obtained during the mission's critical encounter phase. This provided five times as many pictures and much greater high-resolution coverage than would otherwise be possible. The high data rate capability turned out to be crucial to the success of the second encounter, because the tape recorder failed after the first flyby.

Scientific Payload

Since so little was known of Mercury, the scientific instruments to be flown by Mariner 10 had to be chosen carefully in order to explore the planet as thoroughly as possible. To this end, seven experiments were selected: television imaging, infrared radiometry, ultraviolet spectroscopy, magnetic fields, plasma science, charged particles, and radio science. These experiments would provide data to explore the interior, surface, and near-planet environment of Mercury, and also to obtain data on the atmosphere and space environment of Venus.

The television system consisted of two vidicon cameras, each with an eight-position filter wheel. The vidicons were attached to long focal-length Cassegrain telescopes, which were mounted on the spacecraft scan platform for accurate pointing. These telescopes provided narrow-angle, high-resolution images, and were powerful enough to read fine newsprint from a distance of a quarter of a mile. They were absolutely essential for the study of Mercury's surface.

A principal concern of the Imaging Science Team, particularly the atmospheric scientists who were to study Venus's atmosphere, was the inability of the narrow-angle cameras to image large portions of Venus with one picture around the time of encounter. To study atmospheric circulation it is desirable to take pictures of large portions of the atmosphere over a relatively long period, even if the images are at relatively low resolution. Since this flight would be the first time a spacecraft imaged the atmosphere of Venus, atmospheric scientists desperately wanted such pictures. The budget for this project, however, was extremely tight, and any system to obtain such pictures had to be inexpensive and not interfere with the narrow-angle optics. One evening during an Imaging Science Team meeting in Pasadena, several team members were discussing this problem over cocktails in a local restaurant. During this discussion, Bruce Hapke suggested an optical design (which Verner Suomi and Michael Belton sketched on the back of a cocktail napkin) that could provide wide-angle cameras within the budget. The design consisted of auxiliary optics attached to each camera and could be operated by moving a mirror on the filter wheel to a position in the system's optical path. The next morning this design was presented at the team meeting at

JPL, and eventually it was incorporated into the camera system.

The primary objective of the television experiment was to study the physiography and geology of Mercury's surface, determine accurately its radius, shape, and rotation period, evaluate its photometric properties (the manner in which light is reflected from its surface), and search for possible satellites and color differences on its surface. The cameras would also take pictures of Venus to determine its cloud structure and atmospheric circulation.

The infrared radiometer would measure temperatures on the surface of Mercury and in the clouds of Venus by observing the thermal radiation. Temperature variations could provide information on the thermal properties of Mercury's surface material and could be used to infer surface roughness, size of the particles that make up the surface, and whether or not there were rock outcrops and their size.

One of the mission's primary objectives was to search for an atmosphere on Mercury. Theoretical predictions indicated that the most likely constituents of a presumably tenuous atmosphere would be hydrogen, helium, carbon, oxygen, argon, and neon. Consequently, extreme ultraviolet spectrometers were designed to detect these elements; one was placed on the scan platform and the other mounted on the spacecraft body. They would also perform measurements of the Venus atmosphere and gauge the distribution of hydrogen and helium radiation emanating from outside the Solar System.

The magnetic field experiment consisted of two triaxial fluxgate magnetometers located at different positions along a 6-meter (20-foot) boom extending from the spacecraft. The spacecraft itself generated a magnetic field, so it was necessary to place the sensors at different distances from the spacecraft to measure this field and then to subtract it from the interplanetary field and any magnetic field associated with Mercury. This procedure would permit the accurate measurement of a very weak magnetic field that might be associated with the planet.

To understand the interaction of the solar wind with Mercury, it was necessary to observe the velocity and directional distribution of positive ions and electrons in the wind. Two plasma detectors were therefore located on a motor-driven platform attached to the spacecraft body. This experiment would show whether the solar wind interacted with Mercury in a similar manner as with the Moon, or whether a Mercurian magnetic field strongly influenced its distribution. Thus, the plasma detectors would strongly complement the measurements of any planetary magnetic field measured by the magnetometer experiment.

The charged particle experiment would observe high-energy-charged particles—atomic nuclei—over a wide range of energies and atomic numbers. The objectives of the experiment were to determine the effects of the Sun's extended atmosphere (heliosphere) on cosmic rays entering the Solar System from elsewhere in the galaxy, and to search for charged particles in the vicinity of Mercury.

Finally, the radio waves emitted by the spacecraft would be mathematically analyzed to determine the gravitational effects of Mercury on the predicted trajectory of the spacecraft. In this way, it would be possible to accurately measure Mercury's mass. These data would provide a means of accurately determining Mercury's density and, hence, estimates of its internal constitution and structure. Gases in an atmosphere refract and scatter a radio signal, and by measuring these effects it is possible to calculate atmospheric pressures and temperatures. An occultation experiment would observe changes in the radio waves as they moved through the atmospheres of Venus and Mercury when Mariner 10 passed behind the planets as viewed from Earth.

The Flight

After numerous tests of the spacecraft and science instruments under the hostile conditions expected on the Venus/Mercury mission, all was ready for this epic mission of ex-

At the moment the launch window opended, Mariner 10 was sent on its journey to Mercury on November 2, 1973, at 9:45 p.m. Pacific Standard Time. (Courtesy Jet Propulsion Laboratory.)

This picture of Earth was taken November 6, 1973, by Mariner 10 from a distance of 1 million miles. Most of the image shows the eastern Pacific ocean. This was the first time our planet had been photographed from farther away than the Moon. (Courtesy NASA.)

ploration. The spacecraft had to be launched on November 3, 1973, during a short 1½-hour period. At 12:45 a.m. Eastern Standard Time, Mariner 10 was sent aloft from Launch Complex 36B at Cape Canaveral.

For the first time on any planetary mission, the remote sensing science instruments were turned on quite soon after launch. The purpose was to calibrate them in the well-known environment of the Earth-Moon system. Television pictures of Earth were taken for comparisons with Venus, and pictures of the Moon were taken for comparison with Mercury. At this time the first of many problems to plague this historic mission occurred. Heaters designed to hold the television optics at temperatures of 4° to 15° C failed to operate. It was feared that the temperatures would drop to a level low enough to affect the sensitive optics and distort the images. Fortunately, however, the temperature stabilized at an acceptable level and the

En route to Venus and Mercury, Mariner 10 took this photomosaic of the Moon on November 3, 1973, to test the performance of the television system before the flyby of Mercury. It shows a portion of both the front and far sides of the Moon. Mare Crisium is the circular dark area at the right center. (Courtesy NASA.)

cameras maintained their sharp focus. Pictures of Earth showed complex cloud patterns in about the same detail expected at Venus. They could provide valuable comparisons with the Venus clouds. The spacecraft's trajectory took it over the Moon's north pole, where pictures provided the basis for subsequently improving the lunar cartographic network and extending it more accurately to the far side—a prelude to a similar application for constructing maps of Mercury. Mariner 10 images of some areas on the Moon are still our best source of information for these regions. Plasma, ultraviolet, and magnetic measurements were also made within the Earth-Moon system.

On November 13 the first midcourse correction was successfully executed, and by November 28 it was known that a second correction would be necessary to achieve the required trajectory past Venus. However, the mission plan had always included two maneuvers before reaching Venus. By this time, the launch window had closed and the backup spacecraft, waiting in case Mariner 10 failed, could not be launched. Mariner 10 was now competely alone. If it experienced a catastrophic failure after this time, the exploration of Mercury could not take place again until more than a decade had passed.

To the shock of scientists, engineers, and operations personnel, Mariner 10 began experiencing serious problems just after the launch window closed. When the gyros were turned on to roll the spacecraft through a calibration maneuver, the flight data system, which kept track of spacecraft events, automatically reset itself to zero. Although this was not a serious problem, it suggested something might be wrong with the power system. Then on Christmas day, part of the high-gain antenna's feed system failed and caused a significant drop in the signal power emitted by the antenna. Testing indicated that a joint in the feed system may have cracked due to temperature changes. If this problem persisted, no real-time television images could be transmitted, and most of the best pictures planned for the Mercury exploration would be lost. The antenna healed itself and then failed and healed itself again two more times between December 25, 1973, and January 6, 1974.

When the gyros were turned on for another roll calibration maneuver, the flight data system did not reset itself as it had done before. The spacecraft appeared to be be-

The Mariner 10 trajectory past Venus allowed the spacecraft to measure various aspects of the planet's atmosphere. The infrared radiometer trace measured the temperature of the cloud tops on the night and day sides, the ultraviolet spectrometer (striped pattern on right) detected atmospheric constituents, and the television cameras took pictures of the cloud structure. The occultation experiment enabled Mariner 10's radio signals to penetrate the atmosphere. Changes to these signals permitted measurements of atmospheric temperatures and the identification of cloud layers at different levels above the surface of Venus. (Courtesy Jet Propulsion Laboratory.)

having neurotically. Then on January 8, the spacecraft automatically and irreversibly switched from its main power system to its backup system. If the backup failed, the mission would be over. From this point on, extreme care was taken in changing the power status of the spacecraft.

Finally, some good news was reported. On January 21 the second midcourse correction was successfully completed. This maneuver was required to make Mariner 10 fly through a 400-kilometer-diameter area about 16,000 kilometers to the right and in front of Venus, as seen from the approaching spacecraft, or Mariner 10 would not continue on to its rendezvous with Mercury. Analysis of the trajectory showed that Mariner 10 would fly within 27 kilometers of the aim point—a magnificent achievement comparable to hitting a dime with a bullet fired from a

This photomosaic of Venus, taken through the ultraviolet filters, shows the structure of upper-level clouds in the atmosphere. (Courtesy Jet Propulsion Laboratory.)

distance of about 12 kilometers. At this point, all the science equipment was working well, and even the heaters for the television cameras came back on by themselves.

On January 28 another near-calamity struck. During a series of spacecraft roll calibration maneuvers, a gyro-induced instability caused the expulsion of attitude control gas at a dangerously high rate. Without this gas it would not be possible to keep the antennas pointed at Earth, and we would lose contact with the spacecraft forever. Before the problem could be corrected, about 16 percent of the gas was lost.

Despite these problems, Mariner 10 made its closest approach to Venus at about 10 a.m. Pacific Standard Time on February 5, 1974. It took more than 4,000 pictures of Venus's atmospheric structure and circulation between February 5 and February 13, and acquired a wealth of new information about its atmosphere and environment. The ailing spacecraft now headed for its primary target, Mercury, which it would encounter forty-three days later.

But before it could reach Mercury, another midcourse correction was necessary.

Mariner 10's troubles were still not over. On February 18 the spacecraft lost celestial reference on the star Canopus. Apparently its star tracker had locked onto a small particle that had drifted off the spacecraft. By the time Mariner 10 reacquired Conopus, the gyros had been on for an hour and forty-eight minutes, causing more precious attitude control gas to be lost. As the spacecraft neared the Sun it became hotter, and more particles drifted from the spacecraft, causing the star tracker to lose Canopus lock frequently. More gas was lost. In response, the operations team on Earth devised an ingenious method of conserving attitude control gas, called "solar sailing." By differentially tilting the solar panels to use solar photon pressure on the panels in a controlled fashion, it proved possible to significantly reduce the amount of gas needed for a celestially controlled cruise mode. This and other techniques conserved enough gas to accomplish two subsequent Mercury encounters.

The trajectory past Mercury had been carefully chosen to ensure that the best possible science data could be gathered, and also to allow the spacecraft to return to the planet six months later. This plan required a flyby on the night side at an altitude of about 900 kilometers above the surface. Due to the precise aiming at Venus, only one more midcourse correction was needed to change the flyby from the planet's sunlit side to its dark side. Because the planned trajectory correction maneuver would have caused the loss of too much precious gas in gyro oscillations, it was decided to execute a Sun-line maneuver that would not require the gyros. In mid-March Mariner 10's position and orientation would be such that its rocket engine could be fired toward the Sun without having to roll or pitch the spacecraft. By applying the proper amount and direction of thrust at the right time, the spacecraft would be pushed slightly away from the Sun to fly by the dark side of Mercury. On March 16 the maneuver was suc-

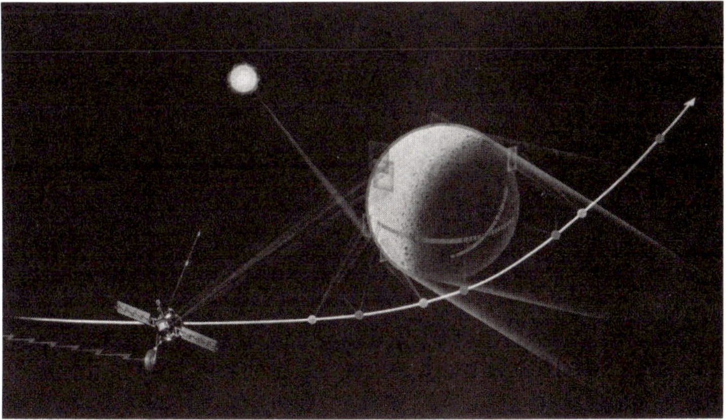

The first flyby of Mercury took place on the planet's night side. This illustration shows Mariner 10's flight path by Mercury and some of the picture footprints and infrared traces on the planet's surface. (Courtesy Jet Propulsion Laboratory.)

cessfully completed, but the flyby would be 200 kilometers closer to Mercury than planned. Since this still satisfied the requirements for science and a Mercury return, no additional maneuvers were planned.

The day after the maneuver, the nonimaging science experiments were turned on in preparation for the encounter. Now some truly good news cheered project personnel, particularly the Television Science Team. The high-gain antenna had miraculously recovered and was able to emit its signal at full strength. Apparently the crack in the antenna feed had healed itself when the temperature rose as the spacecraft approached the Sun. Now most pictures could be taken in real time. The high-resolution coverage originally planned could be accomplished.

The First Encounter

The first television pictures of Mercury were taken on March 24 from a distance of 5.3 million kilometers. They were initially about the same quality as pictures obtained

As Mariner 10 approached Mercury, more details could be observed. The picture at far left, taken March 24, 1974, from a distance of about 5.4 million kilometers, shows about the same amount of detail as the best Earth-based photographs. The right-hand photograph, taken five days later from a distance of about 1 million kilometers, shows Mercury's cratered surface.

from Earth, but as Mariner 10 neared Mercury the images showed a heavily cratered surface superficially resembling the Moon's. The picture-taking sequence called for a series of photographic mosaics to be taken of the half-lit hemisphere as the spacecraft approached the planet. Near closest approach a series of individual high-resolution pictures would be taken near the terminator (the line that separates day and night).

Mariner 10 reached closest approach to Mercury at 1:47 p.m. Pacific Daylight Time on March 29, 1974. For a short time around closest approach, the Earth was occulted by Mercury, cutting off the spacecraft radio signal. All science data, including the highest-resolution television images, were placed on the tape recorder for later playback. As the spacecraft receded from Mercury, a series of photographic mosaics similar to those obtained on approach were taken of the other side of the half-lit planet.

The pictures revealed a heavily cratered surface similar to the lunar highlands. An enormous impact basin about 1,300 kilometers in diameter was revealed half illumi-

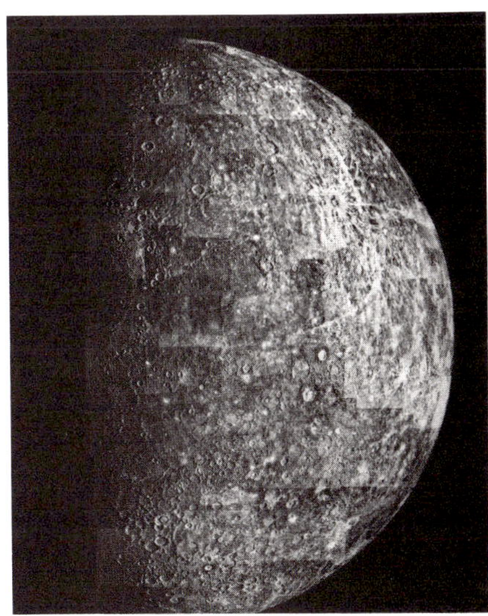

Two photomosaics of Mercury were produced from medium-resolution pictures taken on the first encounter. The one on the left shows the incoming side as viewed by Mariner 10 and the other shows the outgoing side. (Courtesy NASA.)

nated at the terminator. There were also large expanses of smooth, lightly cratered plains that resembled the Moon's maria. All these surface characteristics were similar to the now-familiar features seen on the Moon. There were, however, several aspects of Mercury's surface that differed significantly from the Moon's. Long, sinuous cliffs or scarps traversed the surface for hundreds of kilometers and appeared to occur almost everywhere. The heavily cratered regions contained large areas of moderately cratered plains interspersed among clusters of craters. A large region of hilly and lineated ground—nicknamed the "weird terrain"—was discovered on the incoming side viewed by Mariner 10. These features would enable scientists to reconstruct a geologic history of Mercury that was similar to the Moon's in some respects, but significantly different in others.

Because Mercury was not thought to possess a significant magnetic field, it was generally believed that its interaction with the solar wind would be quite similar to that of the Moon, where the wind impinges directly on the surface and the satellite causes a cavity in the wind behind it. At the Earth, Jupiter, Saturn, and Uranus, the solar wind is held away from the surface by their magnetic fields. Mariner 10's trajectory would carry it through the anticipated plasma cavity behind Mercury. To the astonishment of scientists monitoring the plasma data telemetered from the spacecraft, nineteen minutes before closest approach the plasma flux suddenly increased and peaked in a manner indicating that Mariner 10 had crossed a bow shock wave. At about the same time, the charged particle experiment detected a violent increase in energetic charged particles, confirming that the spacecraft had crossed a shock wave. Several other unusual peaks in the intensity of charged particles occurred subsequently. They were similar to phenomena observed in the Earth's geomagnetic tail, where oppositely directed field lines meet.

Meanwhile, the magnetometer experiment measured an increase in the magnetic field as the spacecraft approached the planet. The interplanetary field typically has a strength of about 6 gammas (units used to measure magnetic intensity), but at closest approach the magnetic field strength reached 100 gammas. If the rate of increase continued to the surface, Mercury would have a magnetic field of about 200 to 500 gammas. Although this strength is only about 1 percent of Earth's field, it is more than adequate to deflect the solar wind and produce the bow shock observed by the plasma and charged particle experiments. Scientists were both elated and perplexed over this completely unexpected discovery. What could be the source of this magnetic field? Was it internally generated or was it due to electric currents induced in the surface by the solar wind? Another Mercury encounter would be required to answer these questions.

The ultraviolet experiment determined that Mercury possesses virtually no atmosphere, and that its ionosphere is no greater than one hundred-thousandth that of the Earth. The planet does have, however, more helium than the Moon, possibly originating from radioactive decay of uranium and thorium or capture from the solar wind.

The infrared radiometer experiment measured a low temperature of 90° Kelvin (-297° F) on the night side just before dawn, and a high temperature of 460° K (369° F) in late afternoon. When Mercury makes its closest approach to the Sun, however, the temperature range can reach 650° K (1,170° F). This enormous temperature difference between night and day is greater than on any other planet or satellite in the Solar System. The temperature gradient between Mercury's day and night side showed that its surface consists of a light, porous insulating layer of dust similar to that on the Moon. Slight temperature variations on the night side indicated the presence of small rock outcrops, probably due to boulder fields around fresh impact craters.

Radio tracking of Mariner 10 provided an accurate radius of the planet and showed that it is much closer to a perfect sphere than either Earth or Mars. Its mass was measured to an accuracy 100 times greater than previous Earth-based determinations. These values could now provide an accurate density of Mercury, which, in turn, would yield information on its bulk composition and internal structure.

Mariner 10's first encounter with Mercury was an outstanding success that exceeded all expectations. The achievement was particularly noteworthy because of the numerous spacecraft problems that had to be overcome and that had threatened to end the mission before it accomplished its objectives. Mariner 10 provided us with our first close-up glimpse of Mercury and returned thousands of photographs and tens of thousands of nonimaging measurements of its surface and environment. A few hours of spacecraft observations had obtained more data about this poorly known planet than centuries of Earth-based observations. They revealed a planet with a combination of Earth-like and Moon-like characteristics that provided important new information on the evolution of the Solar System.

But this was no time to bask in the glories of this success. A second and possibly a third encounter with Mercury were possible, to add to and complement the knowledge acquired during the first. It was now up to the engineers and operations personnel at JPL to guide the ailing spacecraft to further Mercury flybys.

Mercury II

Several additional trajectory corrections were required to return the spacecraft to Mercury. But Mariner 10 would continue to experience serious problems that threatened to terminate the extended mission. Only two days after the first encounter, while the television cameras were still taking far encounter pictures, the temperature in the power electronics compartment rapidly rose and was accompanied by an additional 90-watt drain on the power system. The spacecraft remained on its backup power system, and if it failed the mission would be over. The operations team therefore turned off the cameras and other power-consuming instruments, and implemented additional techniques to accommodate the stress on the power system. This response seemed to stabilize the system, but other problems followed.

Without command, the tape recorder turned on and off several times and then failed completely. Without it, all science data would have to be sent back in real time during subsequent Mercury encounters. The high-gain antenna, which had experienced previous problems, would need to transmit at full strength or most subsequent television pictures would be lost. To make things worse, the flight data subsystem experienced a failure that terminated many of the engineering data channels. This failure greatly increased the difficulty of nursing the ailing spacecraft around the Sun before encountering Mercury for the second time. Finally, the amount of attitude control gas was now quite low because of the oscillation problems, and therefore its use would have to be drastically reduced below the normal cruise rate if two more encounters were to be achieved. At this point it looked as if subsequent encounters with Mercury would have only a slim chance to succeed.

Despite these seemingly insurmountable problems, the exhausted personnel at JPL Mission Operations managed to keep Mariner 10 operating and to preserve enough attitude control gas to accomplish two more Mercury encounters. Because Mariner 10's orbital period around the Sun was almost exactly twice Mercury's period, and since Mercury's rotation period is in two-thirds resonance with its orbital period, the spacecraft would view exactly the same side of the planet on the second and third encounters as it did on the first.

A serious conflict emerged between the magnetic fields and charged particles experiments and the television imaging experiment over the trajectory past Mercury for the second encounter. The fields and particles experiments required a night-side trajectory close to the planet, to obtain information needed to determine whether the magnetic field was internally or externally produced. The imaging experiment needed data on the south polar region, to link the two sides of Mercury seen on the first encounter and to determine the polar distribution of the lobate scarps in order to decide whether they were produced primarily by planetary despinning or by cooling of the interior. This objective required a day-side trajectory at a relatively large distance from the planet. The imaging trajectory would provide little useful information on the magnetic field and interacting particles, while the fields and particles trajectory would yield no new information on Mercury's surface. This was an extremely difficult con-

The second flyby, on Mercury's day side, permitted the acquisition of images of the southern hemisphere. (Courtesy Jet Propulsion Laboratory.)

flict to resolve. After an agonizing evaluation of both cases by the Science Steering Group, it was decided that the second encounter would take an imaging trajectory, and that a third would be planned primarily for fields and particles. This decision was largely based on the improbability of achieving an adequate imaging trajectory on the third encounter if the second encounter followed a fields and particles trajectory.

Mariner 10's second encounter with Mercury (known as Mercury II) was chosen to take place on the sunlit side at an altitude of 50,000 kilometers over the southern hemisphere, 40 degrees below the equatorial plane. This trajectory permitted pictures to be taken of a previously unimaged region and provided a photographic tie over the southern hemisphere between the sides of the planet viewed during the first encounter. A rather large miss distance was required to completely cover the southern hemisphere with narrow-angle pictures at resolutions between 1 and 3 kilometers. Furthermore, the trajectory

This photomosaic of Mercury's southern hemisphere was taken during the second flyby. (Courtesy Jet Propulsion Laboratory.)

would enable a third and final encounter with Mercury. The second encounter was designed primarily as an imaging flyby to photograph new areas and to provide important geologic and cartographic links between the two sides seen previously, thereby facilitating further geologic interpretation. An added bonus would be the acquisition of stereoscopic coverage by combining Mercury I and Mercury II pictures of the same regions taken at different viewing angles.

After two more successful midcourse trajectory corrections, Mariner 10 encountered Mercury for the second time at 1:59 p.m. Pacific Daylight Time on September 21, 1974. About 360 pictures were returned during the three-

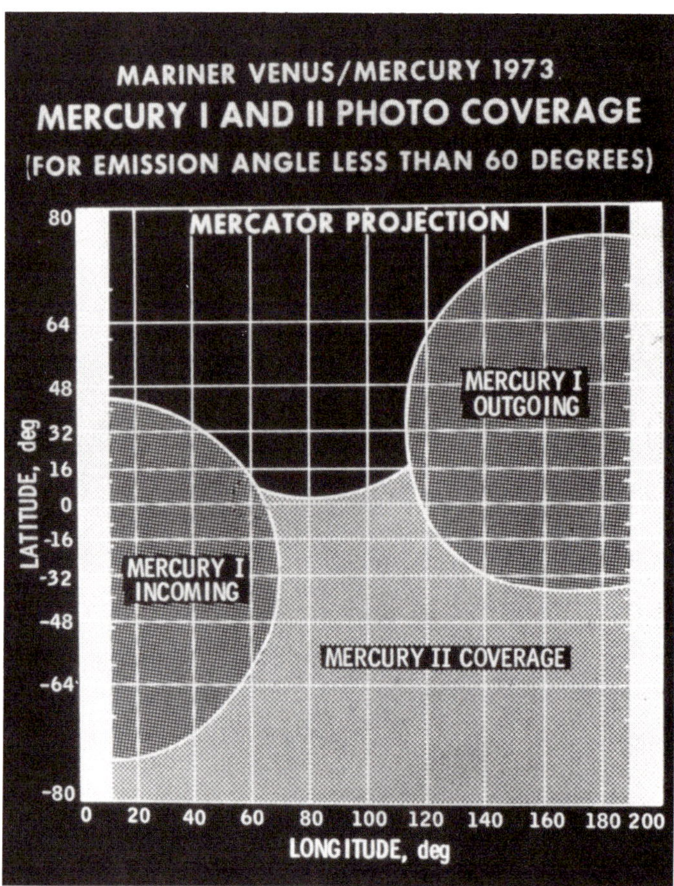

The additional coverage of Mercury obtained by the second encounter joined the two sides of the planet documented on the first encounter. (Courtesy Jet Propulsion Laboratory.)

day encounter sequence. They showed details of the south polar regions never seen before and extended the coverage of Mercury from about 50 to 75 percent of the illuminated hemisphere. The south polar regions revealed a cratered surface similar to that seen on the first encounter. An important observation was the presence of numerous sinuous scarps, first seen on the previous encounter, which indicated that these structures have an extremely widespread distribution, probably on a global scale. This fact would have important implications for interpreting the planet's tectonic history and internal dynamics. The similarity of the terrain in the south polar regions with the terrain on other parts of Mercury increased the confidence of scientists that the half of Mercury viewed by Mariner 10 was fairly representative of the planet as a whole.

The second encounter was historic because it was the first time any spacecraft had returned to its target planet. Furthermore, engineers at the Goldstone Tracking station in the Mojave Desert of California were successful in developing a new technique to obtain relatively noise-free pictures, required because Mariner 10 was at a much greater distance from Earth than at the first encounter and its signal was much weaker. They connected three large antennas by microwave links and operated them as a single large antenna. Without this technique, the quality of the full-frame pictures would have been so degraded that they would have been of little scientific value, or it would have been necessary to transmit only one-quarter of each picture at a lower telemetry rate. In either case, most of the objectives of the second encounter would not have been accomplished. Thanks to the ingenuity of the Goldstone engineers, a full complement of superb television images was obtained in real time.

Although the second flyby was mainly devoted to imaging science, the ultraviolet experiment was able to obtain excellent data. It set even more accurate limits on the density of Mercury's tenuous atmosphere and again observed the emission lines of helium.

Up until this time all interplanetary flights had relied solely on Earth-based radio measurements for navigation. In the second Mariner 10 encounter, a new navigational technique was tested that used the stars as celestial reference points. This navigation method was not unlike that used by ancient mariners to guide them over the vast seas in their explorations of Earth. More than 100 television pictures of star fields were taken to obtain angular measurements between Mercury, the spacecraft, and the stars. The experiment was successful, and it demonstrated that long missions to the outer planets could use this method to navigate spacecraft through the intricate orbits of the outer planets' satellites. Voyagers I and II would later use this technique on their historic explorations of the outer Solar System.

Mercury III

As Mariner 10 began its second orbit around the Sun in preparation for the third encounter (Mercury III), the spacecraft was returned to the cruise mode. The high-gain antenna and solar panels were again used to gather light pressure from solar photons in order to conserve attitude control gas, which was now dangerously low. On October 6 the Canopus star tracker again lost its lock on the star when a bright particle passed through its field of view. The spacecraft went into an uncontrolled roll that could not be corrected before the attitude control gas was depleted below that required to achieve the third encounter.

The situation had now become desperate. What could be done to save the remaining attitude control gas for the crucial third encounter? The operations team decided to abandon roll axis stabilization and permit the spacecraft to slowly roll, the rate controlled by differentially tilting the solar panels. The rates had to be accurately controlled to prevent excessive use of the pitch and yaw jets, and this was made extremely difficult by the earlier loss of many engineering telemetry channels. The continuous rolling of the spacecraft also made navigation much more complicated. Despite these problems, however, three more trajectory correction maneuvers were successfully completed that placed the spacecraft on a path that would take it closer to Mercury (327 kilometers) than any previous planetary flyby.

The measures taken to preserve the attitude control gas seemed to be working, and it looked as if there would be enough to achieve the third encounter. Then, a few days before encounter another problem occurred that nearly ended the mission there and then. While trying to reacquire the reference star Canopus, the spacecraft rolled into a position where the signal strength from the low-gain antenna plummeted to a level that essentially broke communications with Earth. If communications with the spacecraft were not reestablished soon, Mariner 10 would fly by Mercury in utter silence. Only the large 64-meter antennas of the Deep Space Tracking Network were capable of emitting a signal strong enough to command the spacecraft to reacquire Canopus. But these antennas were currently being used to communicate with the Pioneer and Helios spacecraft.

Time was running out, and so a spacecraft emergency was called. To save Mariner 10, the big antenna at Madrid, Spain, was directed to send a command to the spacecraft that, it was hoped, would result in the reacquisition of Canopus and the positioning of the spacecraft to resume normal communications. Even though this was the period

Third Encounter Magnetometer Results

	Time of Observation (PDT) hr:min	
Event	*Predicted*	*Actual*
Cross bow shock	3:31 ± 02	3:31
Cross magnetopause	3:39 ± 01	3:39
Maximum field*	3:49 ± 01	3:49
Recross magnetopause	3:54 ± 01	3:56
Recross bow shock	3:58 ± 02	3:59

*Predicted field strength was 200–500 gamma.
Actual field strength was 400 gamma.

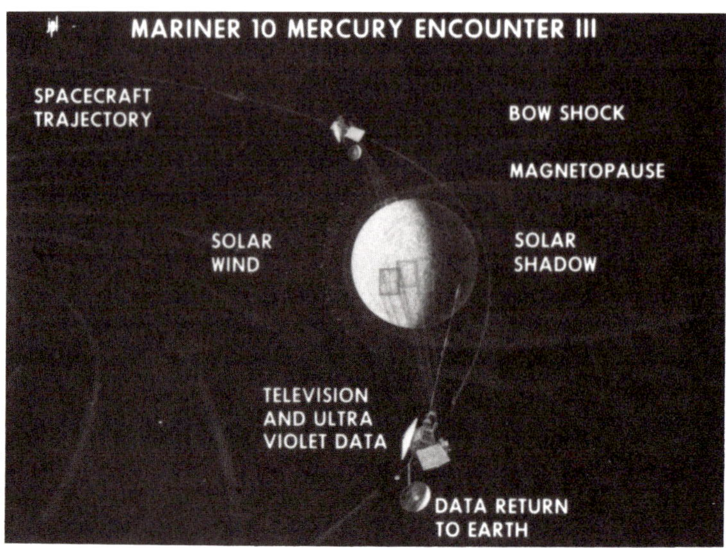

The third and final encounter with Mercury was on the night side, at a distance of only 327 kilometers from the surface. It was designed primarily to answer questions about Mercury's magnetic field, but it also provided high-resolution photographs of the surface. (Courtesy Jet Propulsion Laboratory.)

of maximum scientific interest for the Helios Mission, the Madrid station sent the command, and shortly thereafter, Mariner 10 achieved its correct orientation for the third flyby—just thirty-six hours before its closest approach to Mercury.

The primary goal of Mercury III was to obtain measurements of the magnetic field that would determine whether it was internally generated or produced externally by electric currents induced by the solar wind. If the field was an internally produced, scaled-down version of Earth's magnetic field, then the time expected for events to be observed by Mariner 10 at encounter could be accurately predicted. The actual times that Mariner 10 passed through the bow shock, the magnetopause, and the maximum field strength proved to be almost exactly those pre-

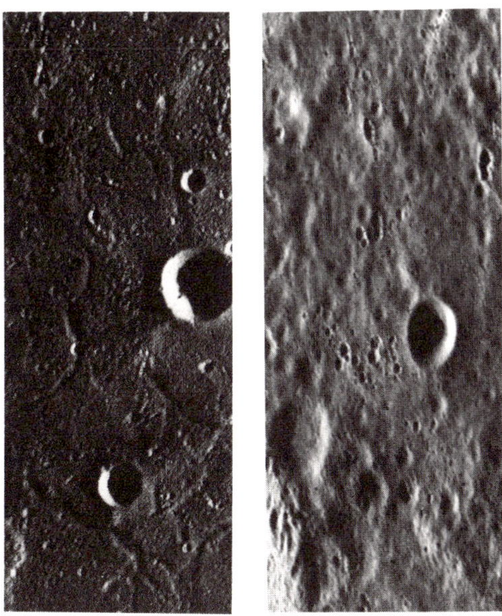

Television pictures taken on the third encounter were limited to quarter frames because of ground-based antenna problems. These pictures included high-resolution images of a portion of the Caloris basin (left) and clusters of sharp secondary craters on the incoming side (right).

dicted. Thus, we have concluded that Mercury's magnetic field is internally generated and similar in form to the Earth's field.

Although the nonimaging science instruments were returning important new results, the imaging science experiment was experiencing serious difficulties. Only the Canberra station of the Deep Space Network was within receiving view of the spacecraft during the third encounter. Earlier, an experimental ultra-low-noise feed was installed so that high-quality, real-time television pictures could be received at the high data rate. Near encounter, however, the feed developed a leak in its cooling system, and the imaging data had to be returned at a much lower data rate. As a result, only a quarter of each picture could be returned in real time. Of course, Mariner 10's tape recorder had failed much earlier, and hence there was no way to store the pictures for later playback to recover the full-frame pictures. The imaging sequence had been planned to return high-resolution images of geologically interesting areas seen during the first encounter. Although much of the picture data was lost, even the strips of high-resolution pictures proved to be important.

The End of a Mission

On March 24, 1975, just one week after the third encounter with Mercury, Mariner 10 ran out of attitude control gas. It began tumbling uncontrollably, and communications were lost forever. The mission was over. On its epic journey of discovery, Mariner 10 had traveled more than a billion miles since it had been launched 506 days earlier. It had transmitted important new information on the atmosphere and space environment of Venus, and provided detailed information about a planet that had baffled scientists for centuries. None of this would have been possible without the herculean efforts and ingenious devices employed by the project personnel to keep the ailing spacecraft alive. Today Mariner 10 continues its endless journey around the Sun, returning every six months to the vicinity of its primary objective, Mercury.

Chapter 3

"Faster Than a Speeding Bullet"

Some of Mercury's most unusual characteristics are its motions. Only Pluto follows a more inclined and elliptical orbit, and no other planet has the peculiar 3:2 resonant relationship between its rotational and orbital periods. Because Mercury is the innermost planet, its orbital speed is greater than that of any other planet. A planet's sidereal period (or year)—the time it takes to travel once around the Sun—depends on its distance from the Sun. This period is described by Kepler's third, or Harmonic, law, which states that a planet's sidereal period in earth years is equal to the square root of its solar distance cubed ($\sqrt{D^3}$), where distance is measured in astronomical units (AU). An astronomical unit is the distance from the Earth to the Sun, or about 150 million kilometers. The average distance from Mercury to the Sun is 0.387 AU (57.9 million kilometers), so Mercury's year is 0.24 earth years, or 87.97 earth days. This means that Mercury travels at an average speed of almost 48 kilometers per second, or about 108,000 miles per hour.

Speed, Eccentricity, and Inclination

Planets travel in elliptical orbits with the Sun at one focus of the ellipse, as first observed by Kepler and stated in his first law. Mercury follows a more elliptical orbit—called its eccentricity—than any other planet except Pluto. The degree of eccentricity is defined as half the distance between the foci divided by the semimajor axis. A perfect

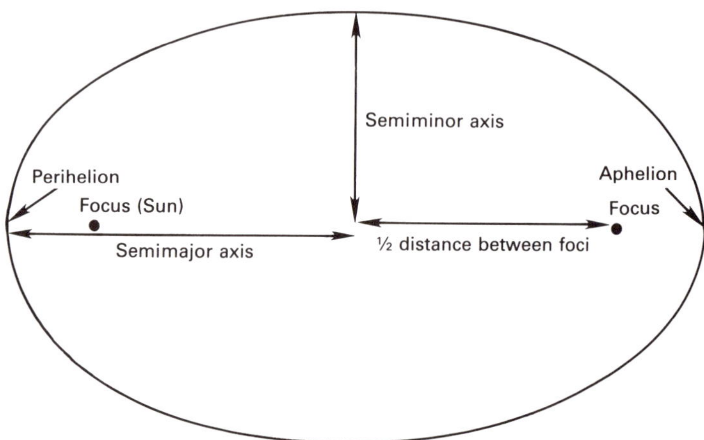

Planets travel on elliptical orbits with the Sun at one focus of the ellipse. The semimajor axis is half the distance of the ellipse's major axis. The eccentricity of the ellipse is half the distance between the foci divided by the semimajor axis.

circle has an eccentricity of zero. Earth's orbit has an eccentricity of 0.0167, which is quite close to circular, but Mercury's eccentricity is 0.205. Therefore, the difference between Mercury's closest point to the Sun (perihelion) and its farthest point (aphelion) is very great. At perihelion, it is at a distance of 0.31 AU from the Sun, but at aphelion, its distance is 0.47 AU—a difference of almost 24 million kilometers.

According to Kepler's second law of planetary motion (the Law of Equal Areas), an imaginary line between the Sun and a planet sweeps out equal areas in equal periods of time. This relationship means that the speed of a planet increases as it approaches the Sun, and decreases as it gets farther away. Because Mercury's eccentricity is so large, it travels much faster at perihelion than at aphelion. At its closest approach to the Sun, Mercury travels at 56.6 kilometers per second, but at its most distant point, it slows to 38.7 kilometers per second. At perihelion, therefore, Mercury travels at the incredible speed of almost 128,000 miles per hour. If an airplane could travel

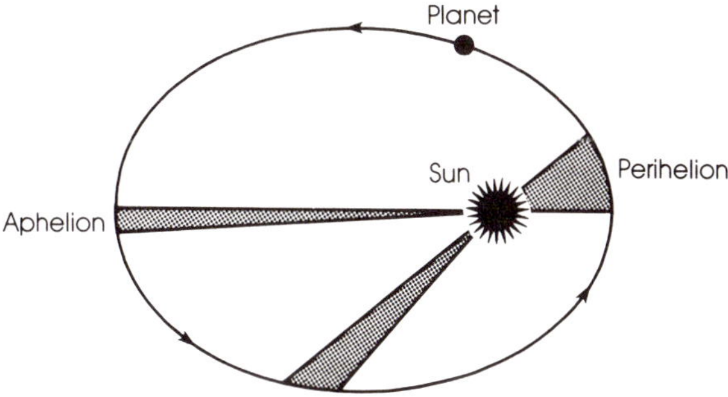

Kepler discovered that a line from the Sun to a planet sweeps out equal areas in equal intervals of time, so that a planet travels faster when closest to the Sun at perihelion and slowest when farthest away at aphelion. (Courtesy Smithsonian Institution Press.)

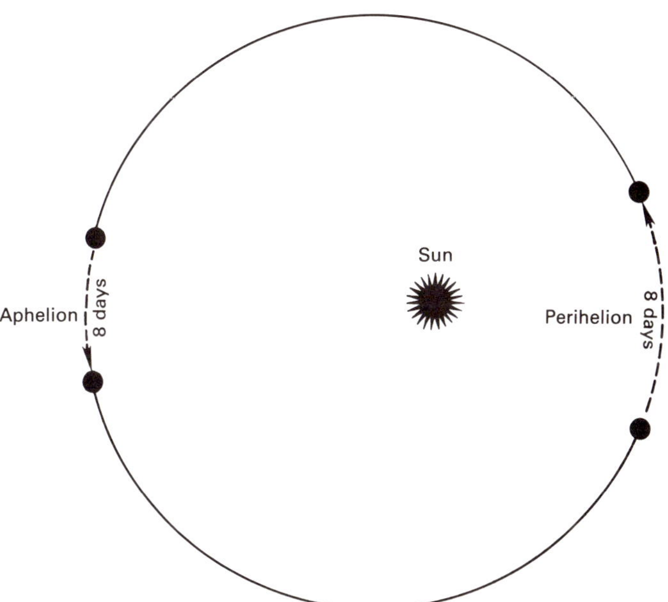

Mercury's orbit around the Sun is more elliptical than any other except Pluto's. Its eccentricity, however, is still small enough that the orbit appears almost circular. The distance Mercury moves in eight days is shown near perihelion and aphelion. Mercury travels much faster near perihelion (56 kilometers per second) than near aphelion (39 kilometers per second).

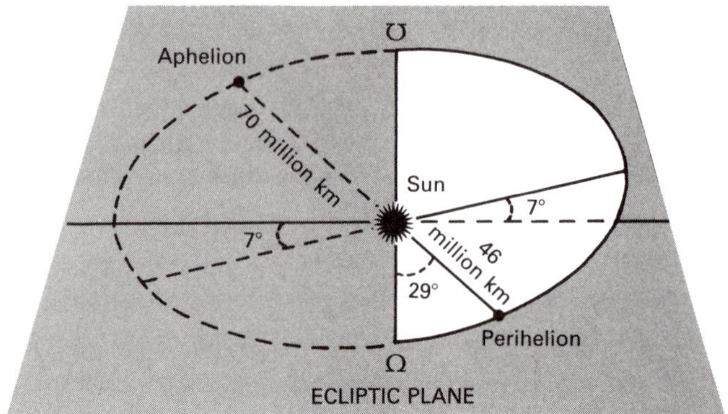

This perspective drawing shows Mercury's orbit with respect to the ecliptic plane (the plane of the Earth's orbit around the Sun). The ascending (☋) and descending (☊) nodes represent the points where Mercury's orbit intersects the ecliptic plane. Mercury's perihelion and aphelion points are 29 degrees from the nodes, and its orbit is inclined 7 degrees (greater than any other planet except Pluto) from the ecliptic plane.

at this speed, it would take less than twelve minutes to circle the Earth.

The inclination of a planet's orbit is a measure of its tilt with respect to the ecliptic plane—the plane of the Earth's orbit around the Sun. Mercury's inclination is 7 degrees, which is exceeded only by the inclination of Pluto's orbit (17 degrees).

Furthermore, unlike the Earth and Mars, Mercury's axis of rotation is perpendicular to its orbital plane. As a result, Mercury has no seasons. Seasons result from the tilt of a planet's axis of rotation in relation to its orbital plane. The Earth's axis is tilted 23.5 degrees to its orbital plane, so that near aphelion the northern hemisphere is tilted toward the Sun and the southern hemisphere is tilted away. At this time, the Earth's northern hemisphere receives more solar radiation per unit area and summer occurs, while at the same time the southern hemisphere is receiving less and is experiencing winter. Six months later, the

Earth has traveled halfway around the Sun and the situation is reversed: winter in the northern hemisphere and summer in the southern hemisphere.

We are not sure why Mercury's eccentricity and inclination are so much larger than those of most other planets. One suggestion is that its orbit has been stretched out and the inclination increased by perturbations caused by Venus's gravitational field. Another possibility is that planetesimals perturbed into the inner Solar System by Jupiter's strong gravitational field collided with Mercury early in its history and knocked it into its present eccentricity and inclination.

Mercury and Relativity

With time, Mercury's elliptical orbit is pulled around the Sun so that the perihelion point changes its position in space. The force that moves the orbit is mostly the gravitational attraction on Mercury by the other planets. In about a century, the perihelion point appears to move around the Sun by approximately 5,600 seconds of an arc (less than 2 degrees). When the effects of the Earth's precession (the changing orientation of the Earth's axis in space) and the gravitational pull of the other planets are subtracted from the 5,600 seconds of arc, there remain 43 seconds of arc unexplained by Newtonian gravitational theory.

Einstein's general theory of relativity predicts that the Sun's mass—745 times the total mass of the planet—warps the nearby space. Since Mercury's orbit is so elliptical, at perihelion it is traveling in space that is more warped than at aphelion. The predicted advance in Mercury's perihelion by moving in this warped space is 43 seconds of arc per century, in agreement with the measured value. This agreement between Mercury's observed perihelion advance and that predicted by relativity theory has been cited as proof of the validity of the general theory of relativity.

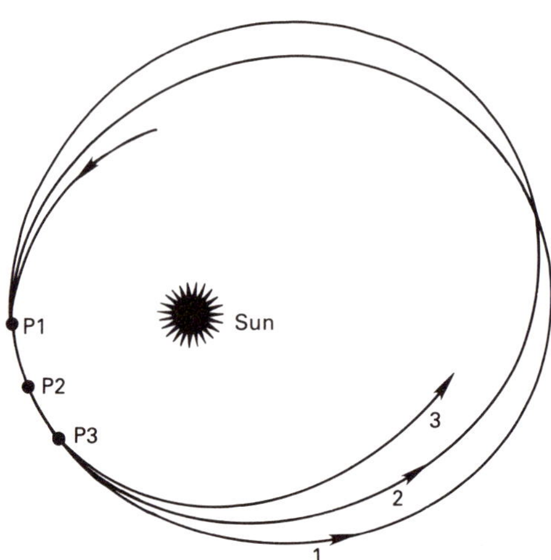

Mercury's perihelion point advances about 5,600 seconds of an arc each century. In this diagram, the time taken for perihelion to advance from P_1 to P_2 (20 degrees) is more than 1,280 years, or about 5,340 orbits. The perihelion advances 43 arc seconds per century more than can be accounted for by perturbations by other planets. This discrepancy can be understood by Einstein's general theory of relativity, because at perihelion Mercury is much deeper in the space warped by the Sun than at aphelion.

Spin-Orbit Coupling

Mercury rotates once on its axis in 58.6 earth days, and it takes 87.9 earth days to complete one circuit around the Sun. Thus, Mercury rotates exactly three times as it circles the Sun twice. This type of relationship between the rotational and orbital period of a planet or satellite is known as spin-orbit coupling. In the case of Mercury, the rotation/orbit relationship is in a 3:2 resonance or commensurability; it makes three rotations for every two orbits. Our own Moon also has a spin-orbit coupling, because it rotates once on its axis during one orbit around the Earth. The Moon, therefore, has a 1:1 resonance.

The 3:2 spin-orbit coupling of Mercury causes a peculiar diurnal (daily) cycle. These combined rotational and or-

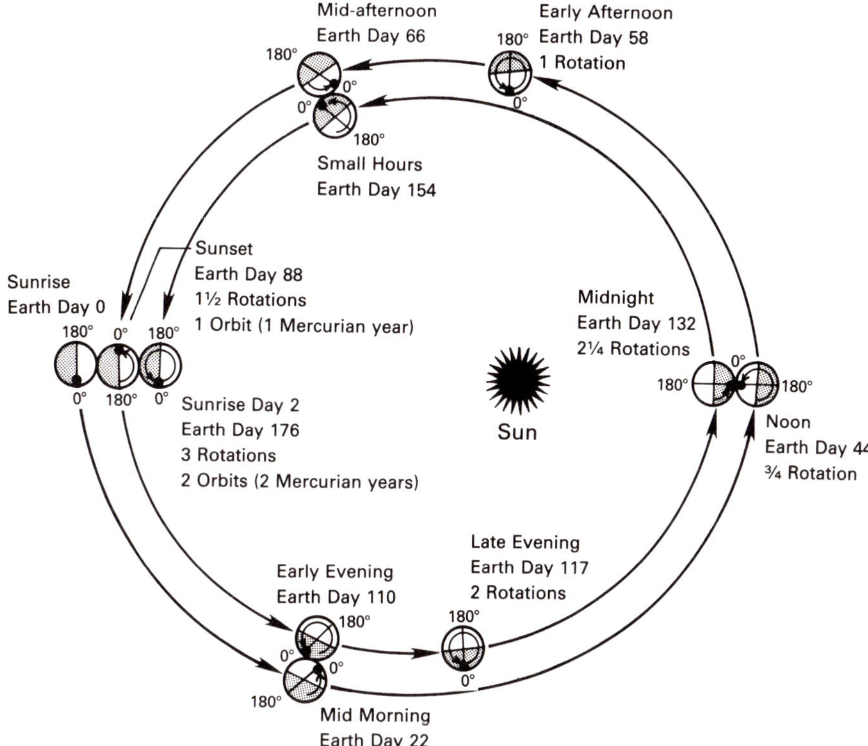

Mercury's 3:2 spin-orbit coupling is shown in this diagram, starting at sunrise on the zero degree meridian. After half an orbit, Mercury has rotated three-quarters of a turn and it is noon at perihelion. After one complete orbit, Mercury has rotated 1½ times, and it is sunset. At the next perihelion passage, Mercury has rotated 2¼ times, and it is midnight. After two orbits, Mercury has rotated three times, and it is again sunrise. Thus, after two orbits, Mercury has rotated on its axis three times and has experienced one solar day.

bital motions result in a Mercurian solar day (sunrise to sunrise) lasting two Mercurian years, or 176 earth days. This is because Mercury's rapid speed around the Sun reduces the effect of its axial rotation. During the time between sunrise and noon, Mercury has completed three-quarters of its axial rotation, but during the same time it has traveled halfway around the Sun. If the planet had remained where it was at sunrise and completed three-quarters of its rotation, it would have been midnight instead of noon.

Another effect of Mercury's 3:2 resonance is that the same hemisphere always faces the Sun at alternate perihelion passages. This pattern occurs because the hemisphere facing the Sun at one perihelion will rotate 1½ times by the next perihelion, placing it directly away from the Sun. But at the following (second) perihelion it will have rotated another 1½ times, placing it directly facing the Sun again. The prime meridian (zero degrees longitude) was chosen to pass through the subsolar point at the first perihelion passage that occurred in 1950. Thus, the hemisphere centered on the zero degree longitude will face the Sun at one perihelion and the opposite hemisphere centered on the 180 degree longitude will face the Sun at the next perihelion; each hemisphere alternately faces and turns away from the Sun every perihelion passage. Conversely, at aphelion the hemisphere centered on the 90 degree meridian will face the Sun, and the opposite hemisphere centered on the 270 degree meridian will face the Sun at the following aphelion. Thus, the hemispheres centered on the zero and 180 degree meridians face the Sun at perihelions, and the hemispheres centered on the 90 and 270 degree meridians face the Sun at aphelions. The subsolar points at zero and 180 degrees longitude are therefore known as "hot poles," because they receive more heat at perihelion, while the subsolar points at 90 and 270 degrees are called "warm poles," because they receive less heat at aphelion.

Still another consequence of the 3:2 resonance is that observers on Mercury would witness peculiar motions of the Sun. During a single perihelion passage, observers at the 90 and 270 degree longitudes would see two sunrises and two sunsets. At perihelion Mercury's orbital speed is so fast compared to its rotational speed that an observer at the 90 degree longitude would witness the Sun rise, hover in the sky, set, and then rise once again. An observer at the 270 degree longitude would see the Sun set, rise again, and then set once more. At the next perihelion the observer at 90 degrees would see the double sunset and

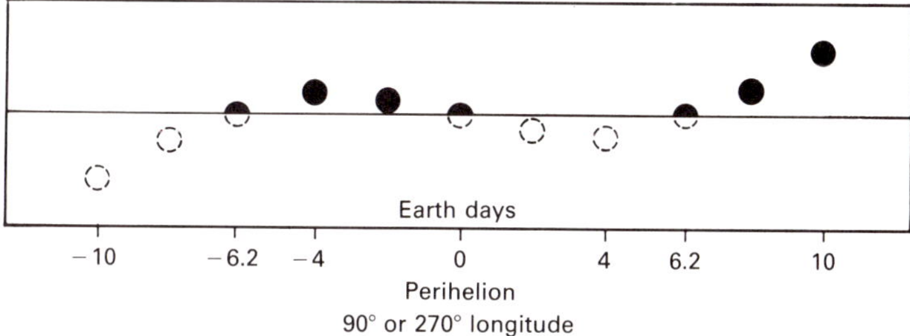

Because of Mercury's rapid orbital speed and slow rotation, a person at its 90 degree or 270 degree longitude would see a double sunrise or double sunset during perihelion passage. About six days before perihelion, the Sun would rise about halfway above the horizon. Four days before perihelion, the Sun would reach about 3.8 degrees above the horizon and then begin to set. At perihelion, the Sun would be half set, and about six days later it would begin to rise again to continue its journey across the sky.

the one at 270 degrees would see the double sunrise. At the zero and 180 degree meridians an observer would see the Sun rise in the east and climb slowly for 1½ earth months until it reached noon. Just before noon the Sun would appear to perform a loop in the sky—slowing down, stopping, backing up, stopping again, and then continuing westward until it set 1½ months later.

Mercury probably did not always rotate with its present period. Just after formation Mercury probably rotated much faster and was subsequently slowed by solar tides. Dynamical studies suggest that Mercury may have had an initial rotation period as short as eight hours. At this rapid rotation rate, centrifugal forces would produce an asymmetrical shape with flattened poles and a bulging equator. The Sun's strong gravitational pull would act on the equatorial bulge to produce tidal friction and slow the planet's rotation period. Eventually it would slow to its present rate and become captured into the 3:2 spin-orbit resonance—a dynamically stable configuration. Some of the energy required to despin the planet would be converted to heat and thereby increase Mercury's internal temperature by about 100° C. As Mercury slowed from a rapid rotation rate and centrifugal forces decreased, it would assume a more spherical shape. In doing so, stresses generated in the planet's outer layers may have been strong enough to fracture and deform the crust (see chapter 8).

Atmosphere

The ability of a planet to retain an atmosphere depends on its mass, temperature, and the type of gases that make up the atmosphere. Large, massive planets, like the Earth and Venus, possess relatively strong gravity fields and, therefore, fairly high escape velocities—the speed required to escape a planet's gravitational influence. If the gases are light enough, the temperature high enough, and the escape velocity low enough, then the atoms and molecules that make up the atmosphere will be accelerated to speeds high enough to escape into space.

Mercury is a relatively small planet with a low escape velocity of about 4 kilometers per second. Because it is so close to the Sun, temperatures can become high enough that even the heavier atmospheric gases will be accelerated to speeds greater than the escape velocity. Thus, any gases that issued from the interior early in Mercury's history would have been rapidly lost to space.

The ultraviolet spectrometer on Mariner 10 measured an atmospheric surface pressure of only about 10^{-10} millibars—about one trillion times less than Earth's. This atmosphere is so tenuous that atoms rarely collide with each other. It constitutes a much better vacuum than can be produced in the best laboratory vacuum chambers on Earth. The atoms identified by the ultraviolet spectrometer were primarily helium and hydrogen, which are probably supplied by the solar wind.

In 1985 Earth-based spectroscopic observations revealed sodium surrounding Mercury. The Mariner 10 instrument did not detect sodium because it was not sensitive to this element. The most abundant element detected by Mariner 10 was helium, with a surface density of only 4.5×10^3 atoms per cubic centimeter. Earth-based observations indicate a sodium abundance near the surface of about 1.5×10^5 atoms per cubic centimeter. This figure indicates that sodium is the most abundant component of Mercury's tenuous atmosphere. The origin of the sodium remains a mystery. It may come from the surface or be vaporized from micrometeorites when they impact the planet. This sodium atmosphere may change in a matter of hours because of its interaction with the Sun's ultraviolet radiation.

Temperatures

Temperatures on Mercury vary enormously. The planet is only about 4.8 million kilometers from the Sun at perihelion, so that on its equator near noon the surface temperature reaches 427° C (about 800° F). This temperature is high enough to melt zinc. But at night, just before sunrise, the temperature plunges to a frigid -183° C, or about 300° below zero on the Fahrenheit scale. This represents a temperature difference between day and night of 610° C or 1,130° F. No other planet or satellite experiences such a wide temperature range. The reasons for these temperature extremes are the intense solar radiation, the lack of a dense insulating atmosphere, and the length of a Mercurian day. Sunrise to sunset lasts nearly three earth months, which gives the surface a long time to heat up. But Mercury's nights are just as long, giving the surface a similar period to cool.

A Day on Mercury

Imagine yourself living on Mercury's equator at the prime (zero degree) meridian. What would your day be like? When the planet is at aphelion, you would awaken shortly before dawn to find that the surface temperature was at

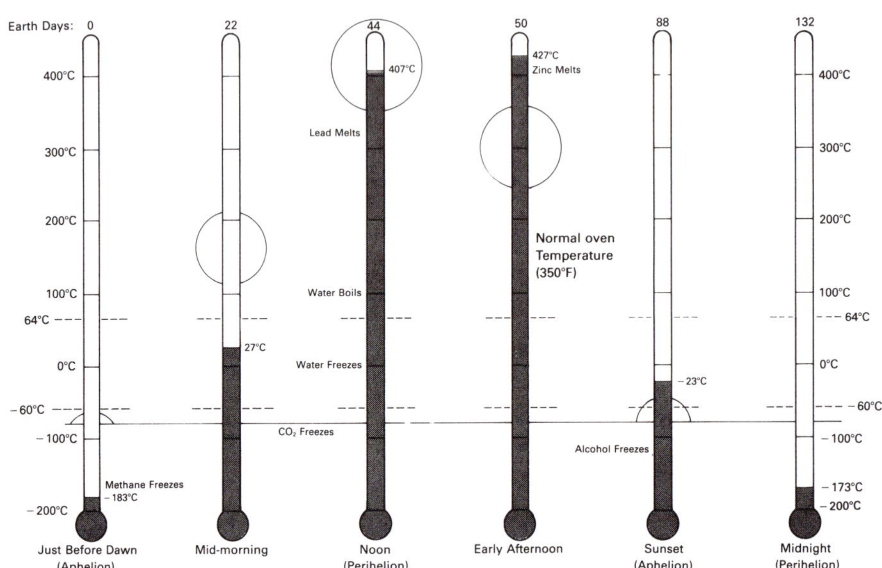

Mercury experiences the most extreme temperature range of any planet or satellite in the Solar System. This diagram shows the change in temperature of Mercury's surface during a solar day. Also shown are the freezing and melting points of several substances and the relative size of the Sun during different times. The temperatures between -60° C and 64° C are the range of air temperatures that can occur on Earth; -60° C is the coldest temperature ever recorded and 64° C is the hottest.

its minimum of -183° C (-300° F). The Sun would rise in the east and slowly move skyward. At this time the Sun's apparent diameter would be twice as large as it is when seen from Earth. But the sky would always be black, because Mercury has essentially no atmosphere. After twenty-two earth days it would be midmorning, and the surface temperature would have risen to a comfortable 27° C (80° F). Twenty-two earth days later it would be noon, the Sun at its zenith performing its curious loop in the sky, signaling that you had reached perihelion and completed half a Mercurian year. The Sun's apparent diameter would now have grown to more than three times its apparent diameter as seen from Earth, and the surface

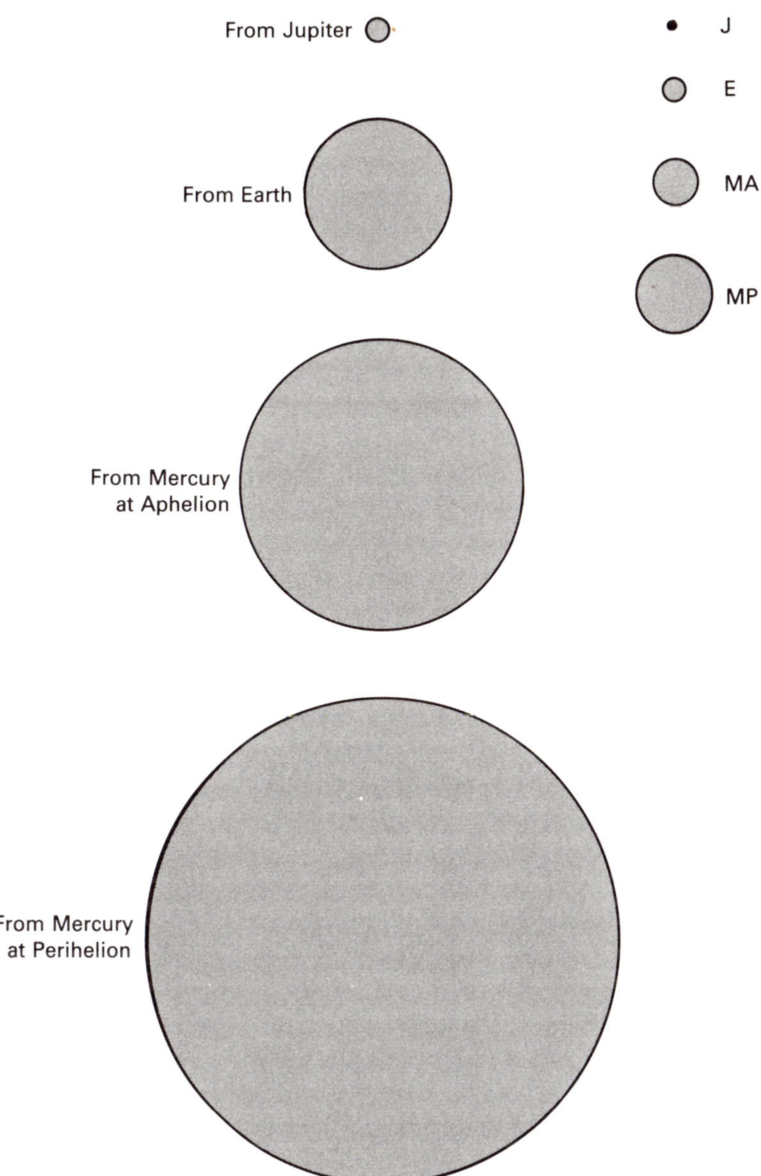

The relative sizes of the Sun as seen from Mercury, Earth, and Jupiter are represented by the circles at the left. Because of Mercury's large orbital eccentricity, the Sun appears much larger at perihelion than at aphelion. When this book is held at a distance of 1 foot, the circles at right are the actual sizes of the Sun as it would be seen with the naked eye from Mercury at perihelion (MP), Mercury at aphelion (MA), Earth (E), and Jupiter (J).

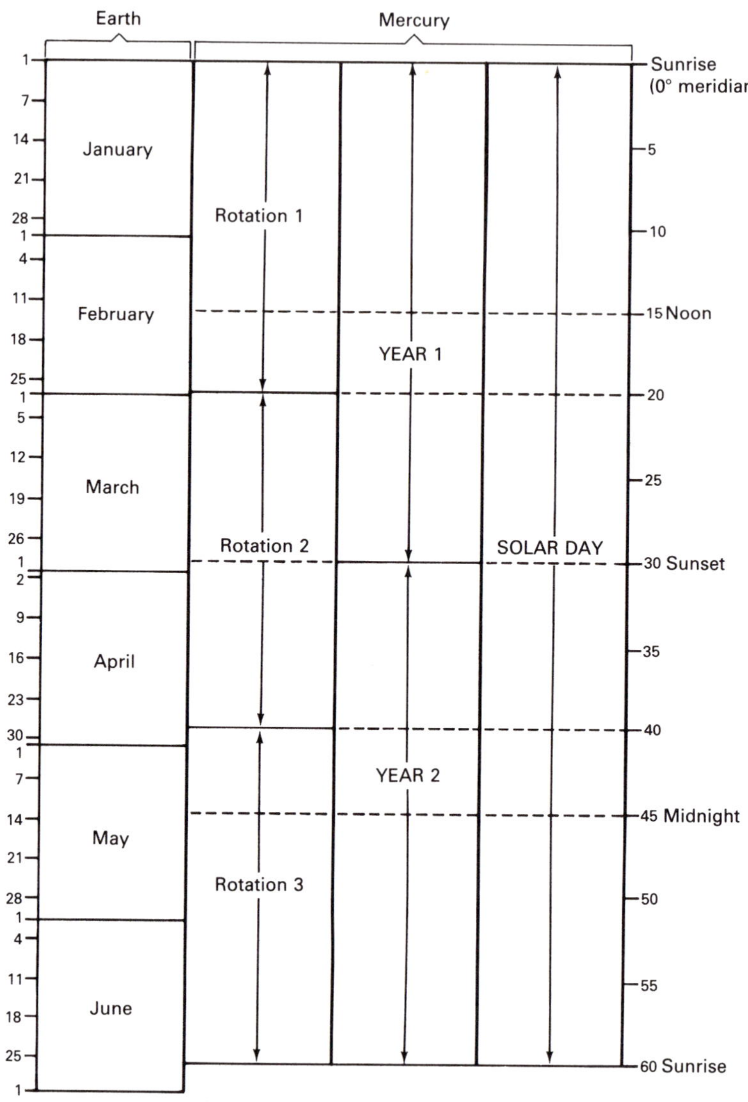

If you lived on Mercury, you might use a calendar such as this one. On this calendar, the Solar day would be the longest unit, lasting almost six Earth months (here arbitrarily divided into 60 units). The day would be divided into two Mercurian years, each of 30-unit duration. Each Mercurian year would be divided into 1½ rotation periods, each rotation period of 20 units' duration. This Mercurian calendar is compared to the first six months of an Earth year.

temperature would have climbed to 407° C (765° F). Six earth days later, in the early afternoon, the surface temperature would reach its maximum of 427° C (800° F). The Sun would continue its westward journey and finally set thirty-eight earth days later, when the surface temperature had fallen to -23° C (-9° F). The sunset would signal that you were again at aphelion and one Mercurian year older.

If you were a night owl, you could stay up until midnight, forty-four earth days later, to observe the constellation Taurus directly overhead, indicating you were again at perihelion. As you retired for the night you would notice that the surface temperature had continued to fall below -70° C (-274° F). When you woke again at sunrise, forty-four earth days later, you would again be at aphelion, another Mercurian year older. During your Mercurian day you would have aged two Mercurian years (about one-half earth year), completed three rotations around your planet's axis, and experienced a surface temperature range of 610° C (1,130° F).

Chapter 4

A Thin-Shelled Iron Sphere

Of the planets, only Pluto is smaller than Mercury. Even three outer planet satellites are as large or larger than Mercury. Callisto, a satellite of Jupiter, is almost exactly the same size as Mercury, while both Ganymede (another satellite of Jupiter) and Titan (a satellite of Saturn) are somewhat larger. Mercury measures only 4,878 kilometers

Mercury is about one-third the diameter of Earth. (Courtesy NASA/ Goddard Space Flight Center.)

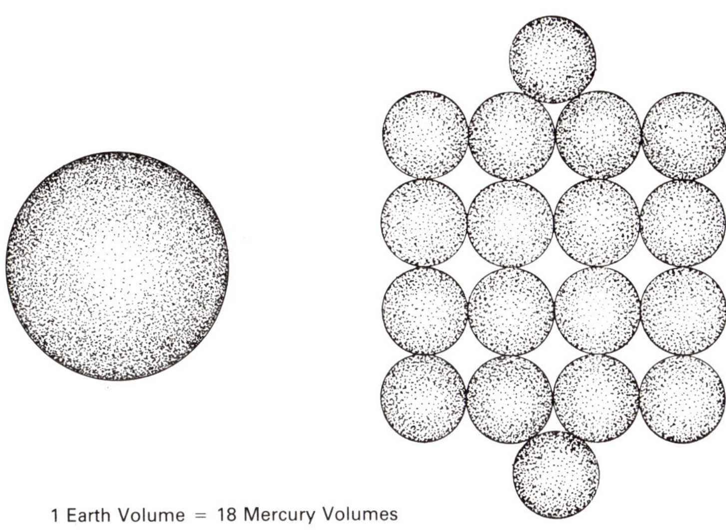

It would take about eighteen Mercurys to equal the volume of Earth.

in diameter, or about a third the size of Earth. Its volume is only 6 percent that of Earth, so that it would take almost eighteen Mercurys to make one Earth.

Size and Mass

Although Mercury is small, it is quite massive compared to its size; therefore, Mercury has the same surface gravity as the larger planet Mars. A planet's surface gravity is a measure of how fast an object is accelerated by its gravity, and is usually measured in centimeters per second per second (abbreviated as cm/sec^2). On Earth, a dropped object will increase its speed by 980 cm/sec each second. The amount by which an object accelerates is determined by the mass and radius of the planet; the larger and more massive a planet, the greater its surface gravity. Although Mercury is 30 percent smaller than Mars, the combination of its large mass and small size results in the same sur-

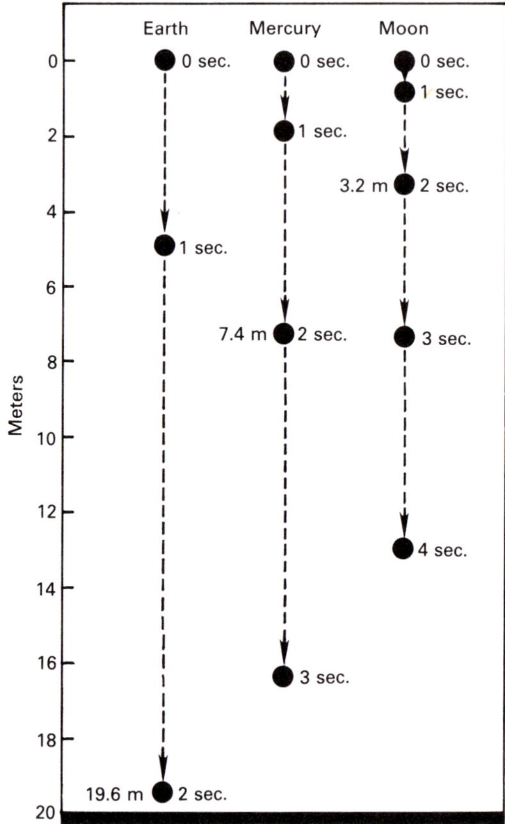

An object dropped on the Earth will accelerate 980 centimeters per second each second, on Mercury 370 cm/sec^2, and on the Moon 162 cm/sec^2. After two seconds, an object on Earth will have fallen 19.6 meters, on Mercury 7.4 meters, and on the Moon only 3.2 meters.

face gravity (370 cm/sec^2). For the same reason, Mercury's gravity field is more than twice as great as the Moon's, although its size is 40 percent larger. Because it contains so much mass in relation to its size, Mercury possesses an unusually high density, only slightly exceeded by the density of Earth.

Density

The density of a planet or satellite reveals something about its gross composition and internal constitution. Density is determined by dividing the mass by the volume, and indicates how much mass is contained in a unit of volume. Density is measured in grams (mass) per cubic centimeter (volume), which is abbreviated g/cm^3.

Planets and satellites are composed of material made up of elements with different atomic masses. Iron is a heavy element with a relatively large mass, while hydrogen and oxygen are relatively light elements with small masses. Therefore, a cubic centimeter of iron has a higher

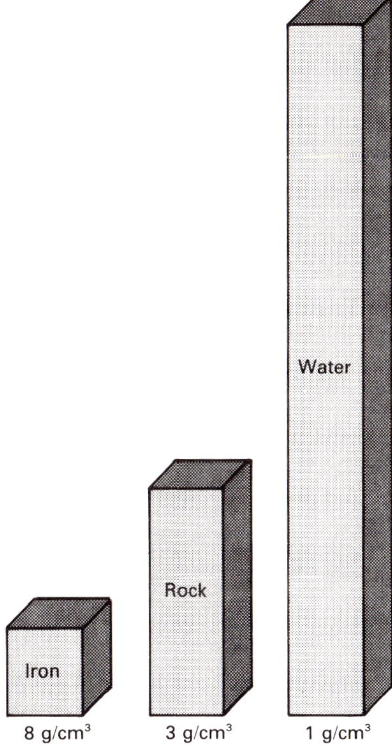

Each of these substances possesses the same amount of mass but different densities.

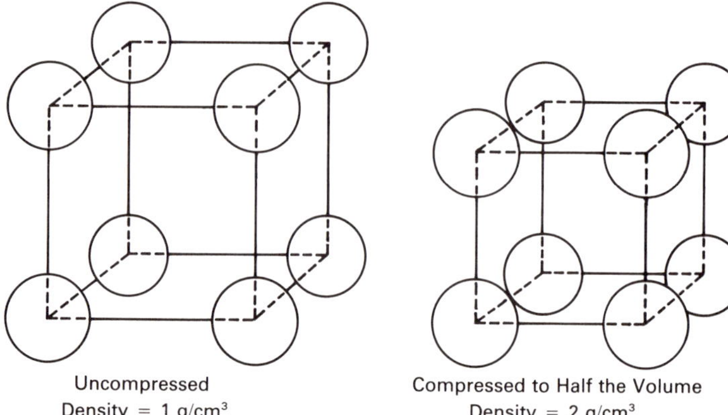

A large amount of pressure can cause the density of a substance to increase. If a material is compressed to half its volume, it still has the same mass but now occupies half the volume. Therefore, the same material has twice the density of its original uncompressed state.

density than a cubic centimeter of water; iron has a density of 7 g/cm^3, and water has a density of 1 g/cm^3. Rocks are composed of various compounds called silicates. They have densities ranging from about 2.6 to 3.3 g/cm^3, depending on their composition. For example, the Hawaiian Islands are composed of a volcanic rock called basalt that is rich in minerals containing iron. Basalt therefore has a relatively high density of about 3.0 g/cm^3. Granite, on the other hand, is poor in iron-bearing minerals and thus has a relatively low density of about 2.7 g/cm^3.

Density can be increased by applying pressure. As a substance is compressed, the atoms of that substance are forced into a smaller volume, and the density increases. For instance, a cubic centimeter of water compressed to half its volume will contain the same number of atoms but now occupying only half the space. Thus, a cubic centimeter of this compressed water will now contain twice the atoms and therefore have a density of 2 g/cm^3 instead of its uncompressed density of 1 g/cm^3. Of course, extreme pressures, found only in the interior of relatively large planets, are capable of causing increases in density.

The Earth as a whole has a density of 5.5 g/cm^3, which is about halfway between the average density of rocks and iron. This density represents an average of the iron core, consisting of 16 percent of the Earth's volume, and the rocky mantle and crust, which make up 84 percent. Because the Earth is so large, pressures in the interior are extremely high. These high pressures compress the atoms of a material into a smaller volume so that they are more closely packed than for the same material near the surface. Thus, a rock or metal deep in the interior of Earth will have a higher density than the same rock or metal found near the surface. When these density or phase changes are taken into account and corrected for the Earth's pressure gradient, the Earth's uncompressed average density is only about 4.0 g/cm^3.

Mercury has a density of 5.4 g/cm^3, which is comparable to that of Earth and Venus (5.2 g/cm^3) but much larger than that of the Moon (3.3 g/cm^3) or Mars (3.9 g/cm^3). However, Mercury is much smaller than Earth and, therefore, pressures in its interior are considerably less than those in Earth's interior. When this factor is taken into account, Mercury's uncompressed average density is still a high 5.3 g/cm^3, which is much larger than Earth's uncompressed density. This must mean that Mercury is composed to a large extent of heavy elements. Iron is the most abundant heavy element in the Solar System and is an important constituent of meteorites and terrestrial planet rocks. Scientists strongly suspect, therefore, that iron is the principal heavy element responsible for Mercury's high density. From this high density we can infer that the planet is composed of about 70 percent by weight of iron and only about 30 percent by weight of rocky material. Mercury thus contains more than twice as much iron per unit volume as any other planet or satellite in the Solar System. The iron is probably concentrated into a core like

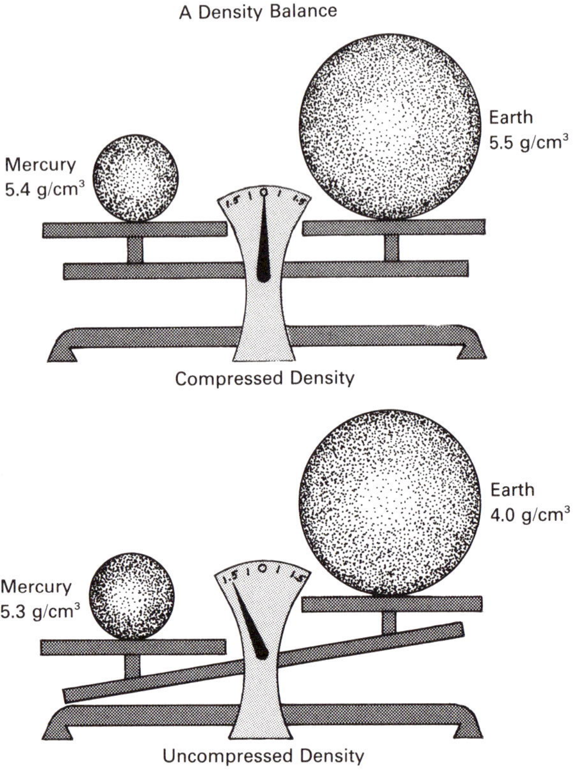

The compressed densities of the Earth and Mercury differ by only about 0.1 g/cm^3, with Earth's slightly greater. The difference between the uncompressed densities, however, is about 1.3 g/cm^3, with Mercury's significantly greater than Earth's.

Earth's, but its size is enormous compared to the diameter of Mercury. The diameter of the core is about 75 percent of Mercury's total diameter and constitutes about 42 percent of its volume. In contrast, Earth's iron core is 54 percent of the total diameter but constitutes only 16 percent of the volume.

Why should Mercury contain such a large proportion of iron compared to other planets and satellites? The reason may relate to its formation so close to the Sun. The Solar System formed from a large cloud of dust and gas called

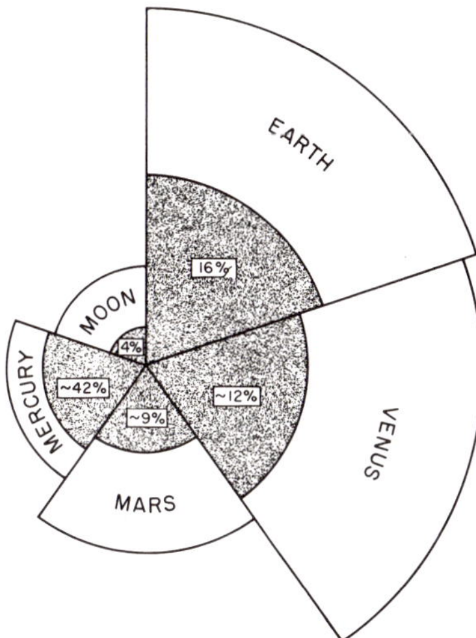

Relative sizes of the Moon and terrestrial planets and their cores. The percent volume of the cores is also shown.

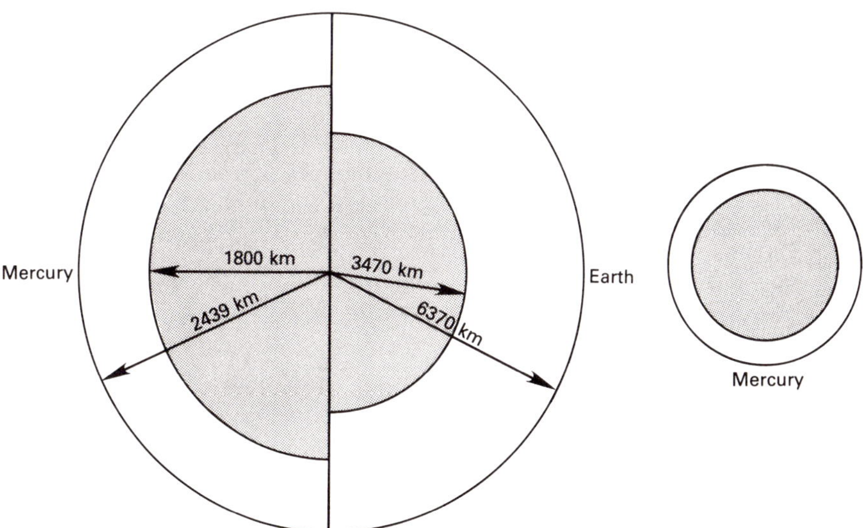

This cutaway view shows Mercury and Earth scaled to the same diameter. It illustrates how large Mercury's core is compared to the planet's total size. Mercury's actual size in relation to the Earth's is indicated by the small disc at the right.

the solar nebula. The center of the nebula was extremely hot (several thousand degrees Centigrade) and the edges were cold (a few tens of degrees Centigrade). Our Sun formed when the hot center of the solar nebula became unstable and collapsed under the influence of gravity. The composition of the solar nebula's dust and gas, from which the planets formed, must have varied greatly with distance from the center because of the large temperature differences. The abundance of volatile, or easily evaporated, elements would increase with respect to the nonvolatile elements at greater distances from the nebula's center. Thus, the terrestrial planets, which formed near the center of the solar nebula, have a much greater abundance of nonvolatile elements such as silicon, aluminum, magnesium, and iron than the outer planets, which have large amounts of volatile elements such as hydrogen and helium. Since Mercury formed in the hottest part of the solar nebula, it would have accumulated from dust richer in metallic iron-nickel alloy than that which formed the more distant terrestrial planets in cooler parts of the nebula. However, Mercury would have to have accreted in an extremely narrow region of this part of the Solar System to have accumulated a large proportion of iron, and even here it could only obtain a density of about 4 g/cm^3. Some other contributing mechanism must have been involved to account for the abnormally high iron content.

Thermal History

Shortly after Mercury formed, heat generated from the decay of radioactive elements such as uranium and thorium, plus heat generated by accretion itself, raised the internal temperature above the melting point. The heavier iron separated from the rocky or silicate material and sank toward the center of the planet. The movement of iron toward the center also produced heat. In Mercury, the large amount of iron moving toward its center would have been great enough to raise the average temperature by about

700° C. Furthermore, Mercury's rotation rate was probably decreasing at this time, which would also have raised the internal temperature by about 100° C. (see chapter 3).

The combined heating by all these processes must have been sufficient to almost completely melt Mercury to an extent at least equal to, and probably greater than, the other terrestrial planets. As material is heated it expands, and the amount of heating experienced by Mercury may have been sufficient to increase its diameter by 30 kilometers or more. In the molten interior, convection currents developed and transferred heat from the interior toward the cooler outer layers, where it was conducted through the solid crust and finally radiated to space. As Mercury lost heat and cooled from this molten state, it contracted. This heating and expansion, and subsequent cooling and contraction, have had profound effects on the surface processes of Mercury (see chapters 7, 8, and 9).

Magnetic Field

Because Mercury is such a small planet, the amount of heat it has lost over the age of the Solar System (4.6 billion years) should have been sufficient to completely solidify the entire planet: core, mantle, and crust. Since the generation of a magnetic field depends on a molten metallic core, scientists long expected that Mercury would not have an appreciable magnetic field, and that the solar wind would interact with the planet in a fashion similar to that of the Moon. Therefore, Mariner 10's discovery of a significant but weak magnetic field came as a complete surprise.

Five planets are known to have appreciable magnetic fields. Of the terrestrial planets, only Earth and Mercury have significant fields. The outer planets of Jupiter, Saturn, and Uranus also have magnetic fields that are stronger than the Earth's and Mercury's.

Earth's magnetic field is dipolar—it has a north magnetic pole and a south magnetic pole, like a bar magnet.

The magnetic axis—the line joining the north and south magnetic poles—does not coincide with Earth's rotational axis but instead is inclined to it by 11.5 degrees. The overall shape of Earth's field also resembles that of a bar magnet: magnetic field lines follow an arcuate pattern between the poles. The strength of a magnetic field is measured in gauss or gammas (1 gauss = 100,000 gammas). The strength of Earth's field at the surface near the equator is about 0.3 gauss or 30,000 gammas.

The solar wind consists of a plasma of charged particles emitted from the Sun that stongly interact with the Earth's magnetic field, compressing the field toward the Earth in a direction away from the Sun. As the high-speed solar wind encounters the magnetic field it forms a bow-shaped shock wave. Between the shock wave and the distorted geomagnetic field is an elongated cavity called the magnetosheath. The inner surface of this elongated cavity is the boundary of Earth's magnetic field and is called the magnetopause. It occurs at a distance of about 11 Earth radii forward of Earth. Encased in the magnetopause is the magnetosphere, which extends for millions of kilometers away from the Sun like the tail of a comet. High-speed solar wind particles and other atoms are captured, trapped, and accelerated in the magnetosphere to form the Van Allen radiation belts. These particles descend along magnetic field lines to collide with atoms high in the Earth's atmosphere, causing them to glow and fluoresce. This process produces the auroral zones and the familiar northern lights.

Mercury's magnetic field is similar to Earth's, but weaker. Its strength at the surface is only about 350 gammas (1 percent of Earth's field strength). Like the Earth's, Mercury's magnetic field is dipolar, with the magnetic axis also inclined about 11 degrees from the rotation axis. It also has the same polarity as the Earth's field; a compass needle will point north. Although Mercury's magnetic field strength is much smaller than Earth's, it is still strong enough to form a bow-shaped shock wave, a mag-

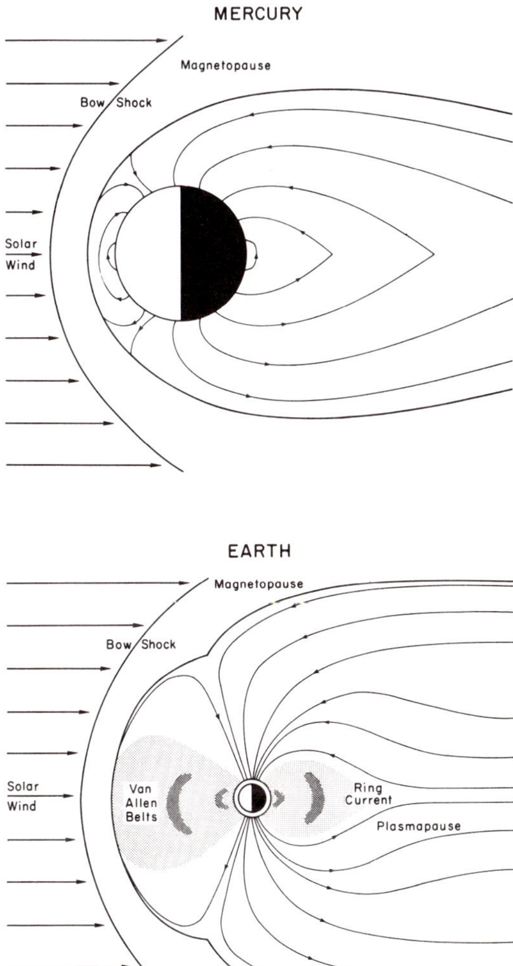

A comparison of the magnetic fields of Earth and Mercury. The drawings are not to scale.

netosheath, and a magnetosphere. Because of the weaker field, however, Mercury encompasses much more of its magnetosphere than Earth. The average distance from the center of the planet to the magnetopause is only 1.5 planetary radii, compared to 11 radii for Earth. As a consequence, charged particles from the solar wind reach Mercury much more easily than they reach Earth, and therefore there are no trapped radiation regions similar to Earth's Van Allen belts.

Planetary magnetic fields are thought to be produced by fluid motions (convection) within electrically conducting molten cores. Fluid motions in such a core generate electrical currents—by electromagnetic induction—that produce the magnetic field. Iron is a good electrical conductor and is therefore capable of producing terrestrial planet magnetic fields. The strong magnetic fields of Jupiter and Saturn, however, are generated by fluid motions in liquid metallic hydrogen, another good electrical conductor.

It is not entirely certain why the Earth and Mercury have maintained their magnetic fields while Venus and Mars have not. At one time, a rapid rotation rate was thought to be essential in producing fluid motions in a molten iron core that, in turn, could generate and maintain a magnetic field. Recent studies, however, indicate that a planet's rotation rate is not as important to the strength of a magnetic field as previously believed. Both Venus and Mars are thought to have molten iron cores, but Mars rotates rapidly (24½ hours) and Venus rotates quite slowly (243 days). Neither planet has a magnetic field. Mercury also rotates slowly (58 days), but it has a magnetic field.

One possible explanation for these apparent discrepancies is that completely fluid cores without intrinsic heat sources may not be able to sustain thermal convection over the age of the Solar System. Instead they cool to a thermally conductive, but still molten, state that cannot maintain a dynamo. Planets that develop solid inner cores may continue to sustain a dynamo because of the release of gravitational energy, heat of solidification, and chemically driven convection that accompany inner core growth. Chemical convection can occur from the release of buoyant material during solidification of an inner core. Venus may not have a significant inner core because of its higher internal temperature and lower central pressure relative to Earth. Mars may not have an inner core either, because its core may have more than 15 percent sulfur, which lowers the melting point and prevents freezing over the age of the Solar System. By contrast, Earth definitely has a solid inner core due to loss of heat and the high central pressure. If Mercury's core contains even a small amount of sulfur, then complete freezing of its core is inhibited and a dynamo can be maintained by chemical convection in a relatively thin fluid shell.

Another explanation for Mercury's magnetic field is that it is a remanent field frozen into an iron-bearing mantle when it solidified from a melt. In this case, the present core would be solid and the observed field would be what remains frozen into the mantle from a magnetic field produced when the core was still molten. There are several problems with this explanation. One is that the amount of iron contained in the mantle to account for the observed field strength seems excessive, considering the extent to which Mercury was probably melted. Another problem is that the temperature of the mantle would have to be low—less than 700° C—or the rocks would not retain their remanent magnetization. Still another problem is that the cooling and solidification of such an enormous iron core should have resulted in a much greater decrease in Mercury's radius than deduced from its surface features (see chapter 8).

Interior Structure and Composition

Our current understanding of Mercury's interior, based on the density and magnetic field, is that it consists of an enormous, at least partly molten, iron core that extends over about 75 percent of Mercury's diameter and occupies about 42 percent of its volume. Overlying the core are a solid mantle and crust of silicates (rocky material), whose combined thickness amounts to only about 600 kilometers. The formation of such a large iron core would lead to extensive melting and crustal deformation, which might be expected to produce surface features unique to Mercury. Although Mercury's surface superficially resembles the Moon's, it does indeed possess geologic characteristics that set it apart from the Moon and other planets. As we shall see, these characteristics are probably the direct or indirect result of the formation of Mercury's large iron core.

Chapter 5

A Pockmarked and Wrinkled Face

Mariner 10 photographed only about 45 percent of Mercury's surface. The resolution of this photography—the smallest detail that can be seen—ranges from about 2 kilometers down to 100 meters in extremely limited areas. Our current coverage of Mercury is roughly equivalent to our knowledge of the Moon in 1965—coverage of one hemisphere at approximately 1–2 kilometers resolution (from Earth-based telescopes) and selected coverage of some areas at somewhat higher resolutions (better than 0.5 kilometers from Ranger). Nevertheless, Mariner 10's images are infinitely better than anything we could ever obtain from Earth, and they have provided new insights into the evolution of Mercury and the other terrestrial planets. In addition, detailed radar altimetry of the equatorial regions of Mercury has been obtained since 1980 by Earth-based radar facilities at Arecibo in Puerto Rico and Goldstone in California. These data have provided important information on the topographic heights and profiles on both imaged and unimaged sides of Mercury.

Mapping Mercury

Maps of planetary surfaces are essential. They show the distribution and orientation of surface features and provide a base upon which a variety of scientific data can be plotted. Planetary maps, therefore, greatly aid data analysis and interpretation.

Photograph of the small crater Hun Kal, which was used to define the 20 degree meridian on Mercury. (Courtesy NASA.)

There are many types of maps, but essential to all is a coordinate or reference system (latitudes and longitudes) by which surface features can be located. Latitudes are small circles centered on the rotation axis, with the equator or zero latitude located 90 degrees from the poles. Longitudes are great circles (meridians) that intersect at the poles of rotation. The location of the prime meridian or zero longitude is completely arbitrary. On Earth, the prime meridian passes through the Greenwich Observatory on the outskirts of London, because it was there that most observations required to define longitudes had been made. Where possible, the prime meridians on other planets and satellites are located on a more practical basis. For instance, the Galilean satellites of Jupiter are all in synchronous rotation, and therefore always keep the same hemisphere turned toward Jupiter. Hence, the prime meridians for these satellites were chosen to coincide with the longitudes passing through their sub-Jupiter points (the point on the hemisphere directly in line with Jupiter).

On Mercury, the prime meridian was selected to coincide with the subsolar longitude when the planet is at perihelion. Because Mercury's rotation period is in a 3:2 commensurability with its orbital period, one hemisphere will face the Sun at one perihelion passage, and the directly opposite hemisphere will face the Sun on the next perihelion passage (see chapter 3). Consequently, there are two perihelion subsolar longitudes 180 degrees apart. Therefore, the International Astronomical Union in 1970 defined the zero degree longitude as the subsolar meridian at the first perihelion after January 1, 1950.

The area containing the prime meridian was not viewed by Mariner 10 because it was 10 degrees into the night side during the three encounters. However, the center of a small, well-defined crater observed on one of the high-resolution images was calculated to lie within half a degree of the 20 degree meridian as defined by the International Astronomical Union's perihelion convention. The center of this crater was therefore selected to exactly coincide with the 20 degree meridian, and to serve as a reference for locating the other longitudes. This 1.5-kilometer crater is called Hun Kal, which is the number twenty in the ancient Mayan language. (The Mayans, the most advanced astronomers in the ancient Americas, used a numbering system based on twenty, rather than the base ten used by Western civilization.) The coordinates of numerous surface features were determined from the Mariner 10 spacecraft ephemeris (position of the spacecraft at various times) and then used to position the latitude-longitude grid relative to the topography. Longitudes are measured from zero degrees to 360 degrees, increasing to the west.

Mercury's surface has been divided into fifteen areas or quadrangles, nine of which have been compiled into shaded relief maps at a scale of 1:5 million. Each map is designated by the letter H (for Hermes, the Greek equiva-

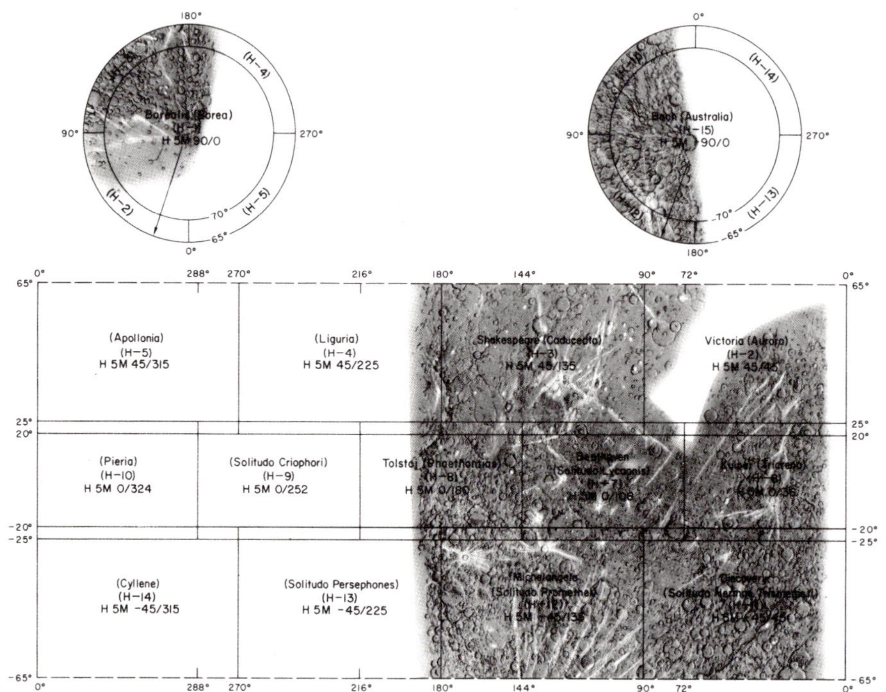

Figure 17 The regions of Mercury as seen by Mariner 10, with the locations of the nine sections of the Atlas indicated

This map shows the distribution and names of the fifteen 1:5 million scale quadrangles. Only nine of these quadrangles were wholly or partly photographed by Mariner 10. About 55 percent of the planet still remains unexplored. (Courtesy U.S. Geological Survey.)

lent of Mercury) followed by a number from 1 to 15. The nine areas viewed by Mariner 10 are further designated by the name of a prominent surface feature contained in the area. For example, the south polar map is called the Bach (H-15) quadrangle, after a large crater in the region.

Names of planet and satellite surface features are approved by a nomenclature committee of the International Astronomical Union. In the case of Mercury, craters have been named after authors, artists, and musicians such as Dickens, Michelangelo, and Beethoven. Two exceptions are Hun Kal and Kuiper, named after Gerard P. Kuiper, a noted University of Arizona astronomer who was a member of the Mariner 10 Imaging Team before his death in December 1973. Valleys (called "valles" from Latin) are named for prominent radio observatories such as Arecibo and Goldstone, while cliffs and scarps (called "rupes" from Latin) are named for ships associated with exploration and scientific research, such as Discovery and Victoria. Two prominent ridges are called Antoniadi and Schiaparelli for the astronomers who first mapped Mercury from Earth-based observations. Plains (called "planitiae" from Latin) are named after the word for the planet Mercury in various languages, and for gods from ancient cultures who had a role similar to that of the Roman god Mercury. Typical names are Odin (Scandinavian) and Tir (Germanic). Borealis Planitia (Northern Plains) and Caloris Planitia (Plains of Heat) are exceptions to this general rule.

A U.S. Geological Survey shaded relief map of Mercury's Shakespeare quadrangle (H-3).

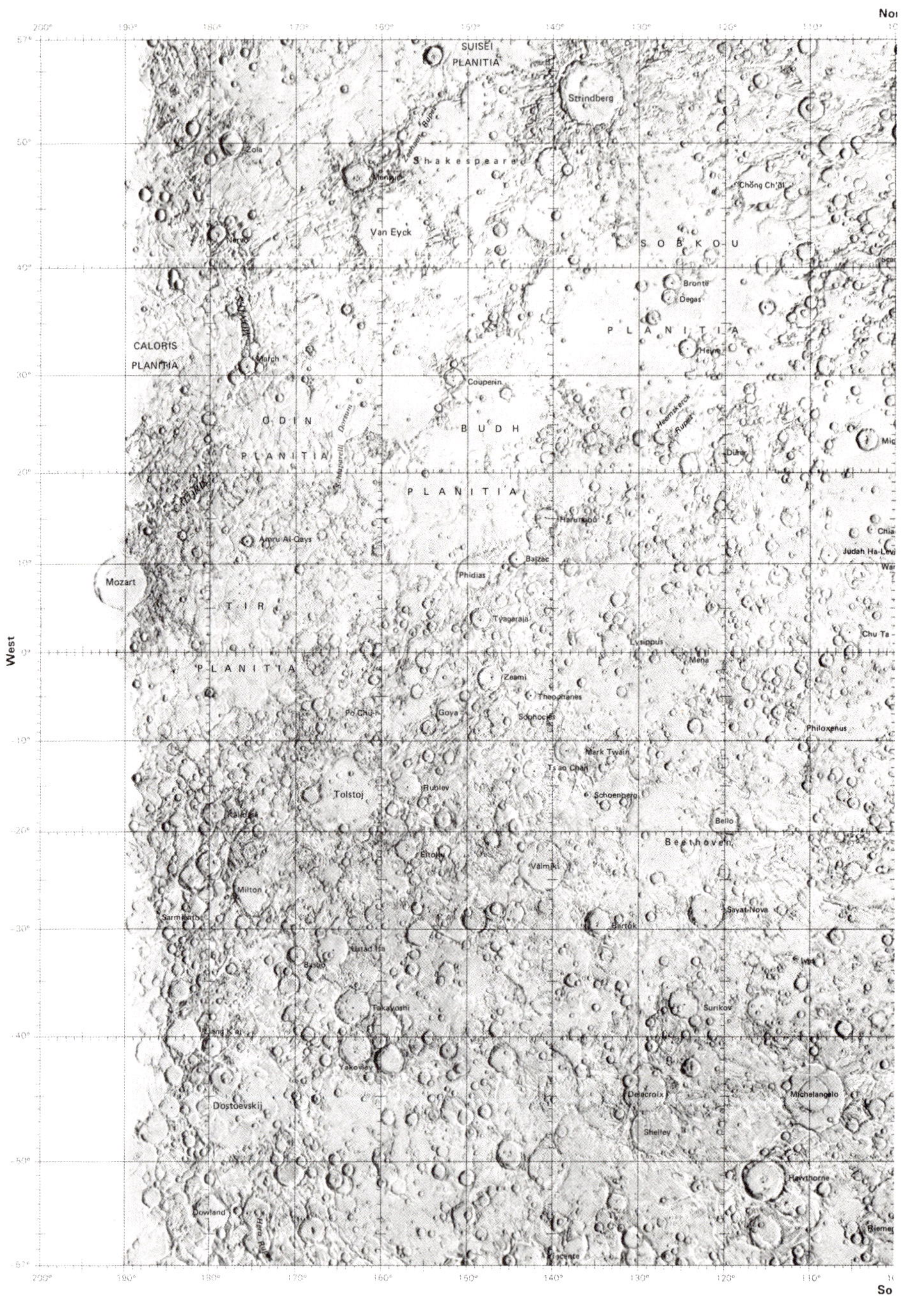

Shaded relief map of the equatorial region of Mercury. (Courtesy of U.S. Geological Survey.)

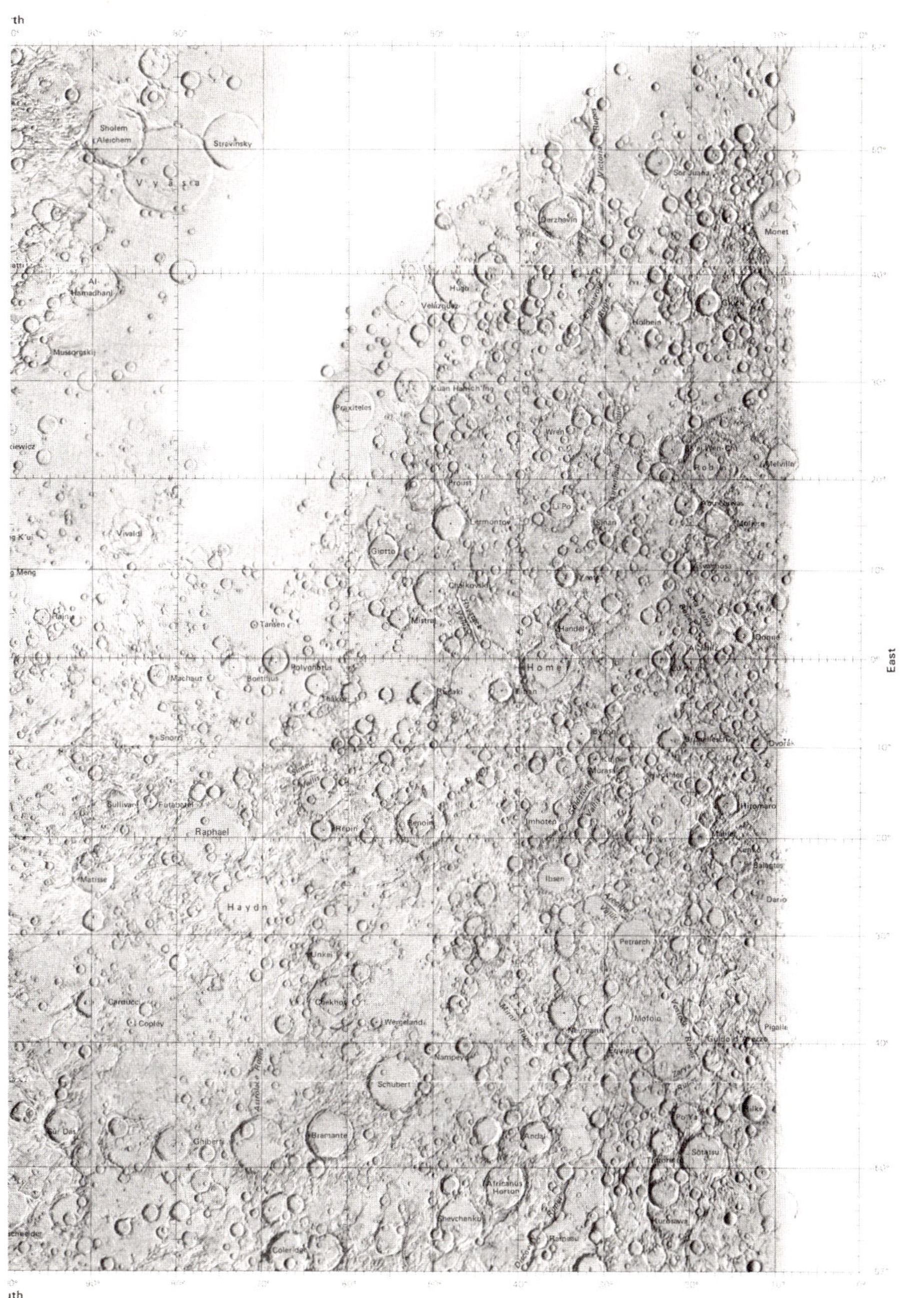
rth
Sholem Aleichem
Stravinsky
V y a s a
Sor Juana
Monet
Derzhavin
Al-Hamadhani
Mussorgskij
Hugo
Velázquez
Holbein
Kuan Han-ch'ing
Praxiteles
Wren
Melville
Proust
Li Po
Lermontov
Sinan
Molière
Vivaldi
Giotto
Chaikovskij
Mistral
Handel
Tansen
Polygnotus
H o m e r
Machaut
Boethius
Snorri
Sullivan
Futabatei
Raphael
Repin
Renoir
Imhotep
Hiroshige
Matisse
H a y d n
Ibsen
Dario
Petrarch
Unkei
Chekhov
Carducci
Copley
Wergeland
Pigalle
Guido d'Arezzo
Nampeyo
Schubert
Ghiberti
Bramante
Andal
Sotatsu
Africanus Horton
Shevchenko
Kurosawa
Coleridge
uth
East

Major Surface Features

As Mariner 10 approached Mercury on its first encounter, it imaged the half-lit hemisphere centered on the prime meridian, and as it departed, it photographed the opposite half-lit hemisphere centered on the 180 degree meridian. On the second encounter, the spacecraft photographed the south polar region, joining the two sides previously im-

Photomosaic of the incoming side of Mercury viewed by Mariner 10. (Courtesy NASA.)

aged on the first encounter. The third encounter concentrated on taking high-resolution pictures of scientifically interesting features discovered on the first encounter. The first two flybys succeeded in imaging almost all of the hemisphere between 10 and 190 degrees longitude. Only about 5 percent of the northern part of this hemisphere was not imaged.

Photomosaic of the outgoing side of Mercury viewed by Mariner 10. Most of the smooth plains are concentrated on this side. (Courtesy NASA.)

At first glance, Mercury looks like the Moon. Most of the surface documented by Mariner 10 consists of heavily cratered uplands grossly similar to the Moon's heavily cratered highlands. Broad expanses of lightly cratered, smooth plains are concentrated on the side of the planet viewed by the departing spacecraft. These plains resemble the lunar maria, the dark, smooth lava plains that occur primarily on the Moon's front side, but they do not display the same contrast in albedo (reflectivity) that is seen between the lunar maria (dark) and highlands (light).

On closer examination, the pictures also show features and terrains not common on the Moon. Large areas of moderately cratered plains are interspersed among heavily cratered upland regions. The very widespread distribution of these plains—called intercrater plains—is not seen on the Moon. Furthermore, Mercury's long, sinuous cliffs, called lobate scarps, cut the surface for hundreds of kilometers. They occur on all types of terrain and all locations imaged by Mariner 10. No other planet or satellite features such a widespread—probably global—distribution of such structures.

The largest feature documented by Mariner 10 is the Caloris basin, an enormous impact crater 1,300 kilometers in diameter that was photographed half lit by the departing spacecraft. The Caloris basin—the Basin of Heat—derives its name from the Latin word "calor," because it lies near one of Mercury's hot poles. It is defined by a ring of discontinuous mountains (Montes Caloris) about 2 kilometers high. The basin's floor consists of smooth plains that have been highly fractured and ridged—unlike any other basin floor yet observed in the Solar System. Beyond the main mountain ring to a distance of about 700 kilometers lie tracts of rough, hummocky plains and lineated terrain caused by deposition of material thrown out by the huge impact that created the basin.

On the other side of Mercury directly opposite the Caloris basin—the Caloris antipodal point—lies a peculiar terrain consisting of hills and valleys that disrupt other

Photomosaic of Mercury's south pole, taken by Mariner 10 on its second encounter. (Courtesy NASA.)

landforms. This region was nicknamed the "weird terrain" by the Mariner 10 Imaging Team, but it was formally termed as "hilly and lineated" terrain. It is found only at this location, antipodal to Caloris, and it probably formed as a result of the impact that created the Caloris basin. A somewhat similar terrain occurs at the antipodal point of the Imbrium basin on the Moon, but it is neither as widespread nor as prominent as it is on Mercury.

Impact craters are the dominant landform on Mercury. They occur to a greater or lesser extent on all surfaces imaged by Mariner 10. Their sizes range from basins as large as Caloris (1,300 kilometers) down to the smallest features Mariner 10 was able to see (100 meters). Much smaller

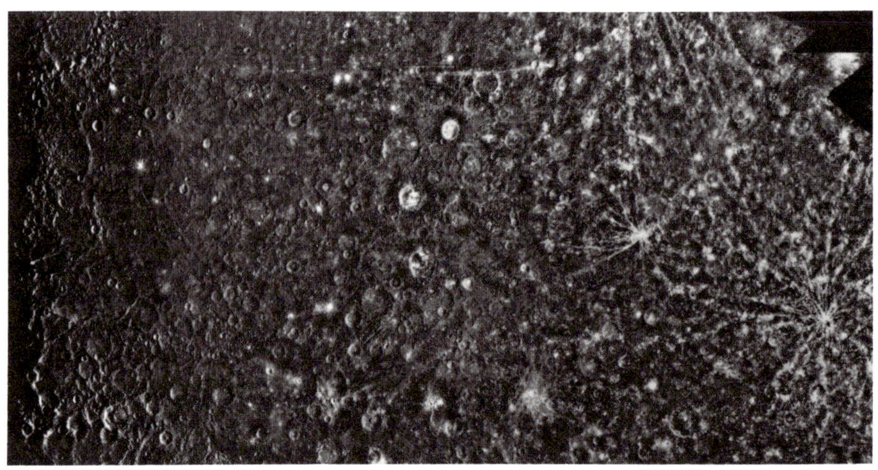

Above and facing: Computer-generated Mercator projection of the equatorial region of Mercury from about 10 to 190 degrees longitude and ±30 degrees latitude. (Jet Propulsion Laboratory.)

craters undoubtedly pepper the surface. The craters are in various states of preservation, from sharp-rimmed with extensive bright rays to highly degraded with low, discontinuous rims.

Plains are Mercury's most common and widely distributed terrain. The younger, lightly cratered, smooth plains are concentrated on the side of the planet viewed by the departing Mariner 10. The largest area of smooth plains fills and surrounds the Caloris basin. Another large tract occurs in a north polar region termed Borealis Planitia. Other areas contain smaller patches of smooth plains, mostly occupying the centers of basins and craters. Older, moderately cratered intercrater plains cover about 45 percent of the heavily cratered uplands viewed by Mariner 10. They are gently rolling surfaces interspersed among heavily cratered regions. Intercrater plains span a range of ages that coincide with a period of heavy bombardment when most of the larger craters were formed.

All these various terrains are cut by lobate scarps forming a unique global fracture system that developed midway through the active part of Mercury's history

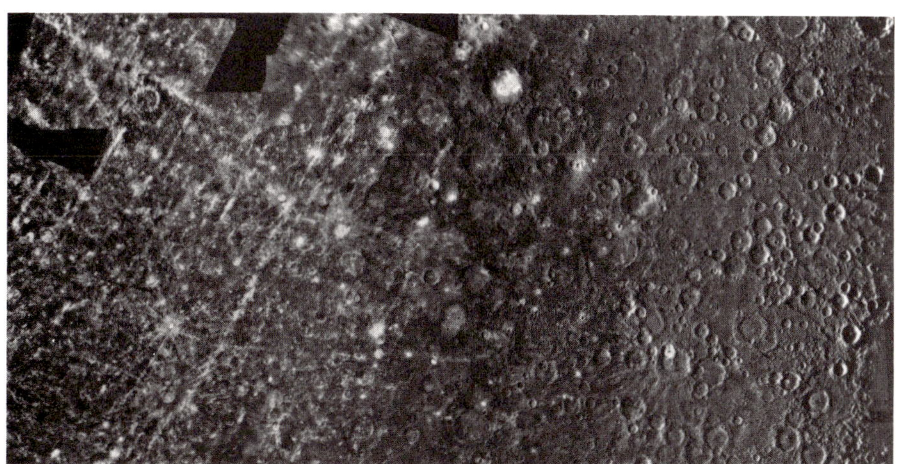

Small-Scale Structure

Although Mariner 10 did not image objects smaller than 100 meters, its instruments revealed some information about the small-scale structure of Mercury's surface. Its infrared radiometer measured the temperature of the surface in various areas on the planet's day and night sides. The rate at which a surface cools measures a property of the material called the thermal inertia—the resistance to changes in temperatures. A bare rock surface will cool slowly and have a high thermal inertia, while a porous, powdery material will cool more rapidly and have a lower thermal inertia. The cooling curve of Mercury's surface during the night yielded low thermal inertia values that are quite similar to those on the Moon. This indicates that Mercury, like the Moon, is covered with a layer of porous silicate dust forming the top of a fragmental deposit termed a regolith. As on the Moon, Mercury's regolith was probably formed by large and small impacts that have pulverized the surface over billions of years. The layer may be a few meters to tens of meters thick, depending on the age of the surface.

In some areas, the Mercurian surface has a relatively high thermal inertia, indicating regions where rock outcrops or boulders are not blanketed by dust. These areas

Comparison of Lunar and Mercurian Normal Albedos

Terrain	*Albedo (%)*
Lunar Maria	6–7 %
Mercury Caloris smooth plains	12–13
Lunar Highlands	10–11
Mercury Highlands (intercrater plains)	16–18
Lunar bright rayed craters	15–16
Mercury bright rayed craters	36–41

may be the sites of relatively fresh impacts that have exposed bare rock and strewn the surrounding surface with boulders. The infrared data are consistent with the photometric, polarization, and radar measurements that also indicate a porous, fragmental surface layer.

Albedo and Color

Until recently, Mercury was thought to be "darker" than the Moon—that is, to have a lower albedo or to reflect less light. Albedo measurements from Mariner 10 images and a reevaluation of Earth-based measurements show that, in fact, the opposite is true. The albedos of various Mercurian terrains are systematically higher than similar lunar terrains. The smooth plains surrounding the Caloris basin are about twice as bright as the morphologically similar lunar maria, and the Mercurian uplands consisting mainly of intercrater plains are more than 60 percent brighter than the lunar highlands. Some of the brightest features on the Moon are the rayed craters, but Mercurian bright rayed craters are more than twice as bright. In fact, the Caloris smooth plains are about as bright as the lunar highlands, while the Mercurian uplands possess about the same brightness as the lunar rayed craters.

These albedo differences may be caused by compositional differences between the Moon and Mercury.

By comparing images photographed through filters that permit the passage of greatly different wavelengths of light, it is possible to detect slight color differences indistinguishable to the unaided eye. In this way, slight color differences have been detected on the surfaces of both the Moon and Mercury. In the case of the Moon, these color differences are found primarily in the maria, where they record differences in the titanium content of the mare lavas. On Mercury the color differences are less than those on the Moon, and they are not found primarily in the smooth plains—the Mercurian equivalent of the lunar maria. Furthermore, the Mercurian ray systems feature a bluish tint, whereas lunar rays have a reddish tint. These differences in the color characteristics again may result from composition differences between the two bodies.

Surface Composition

Little is known about the composition of Mercury's surface. Earth-based spectroscopic measurements may have detected iron-bearing minerals, but their abundance and even their identification remain very uncertain. Minerals containing iron and titanium contribute to darkening rocks, and titanium abundances are primarily responsible for the color differences observed on the Moon. The smaller color differences, the bluish ray systems, and the higher albedos on Mercury, compared to the Moon, have been interpreted to mean that the planet's surface has a lower content of iron and titanium than the Moon's.

The reason for this difference is not clear. One possibility is that Mercury has been more thoroughly differentiated than the Moon. Thermal history calculations indicate that Mercury was more thoroughly melted than the Moon. Under Mercury's stronger gravity field—the same as Mars's—heavy iron-bearing minerals together with titanium, which has an affinity for iron-bearing minerals, would sink faster on Mercury before the outer molten layer could solidify. Thus, these elements may be more concentrated at lower levels on Mercury than on the Moon.

This lack of knowledge concerning the composition of Mercury's outer layers remains a great obstacle in understanding how the planet's crust and mantle evolved. Until this situation is remedied, we can only guess about the chemical evolution of Mercury's rocky outer layers.

Surface Differences between Mercury and the Moon

Although Mercury's surface resembles the Moon's in certain respects, in others it is quite different. In three important respects Mercury's surface is like the Moon's: Mercury's ancient, heavily cratered uplands are similar to the ancient, heavily cratered lunar highlands. Furthermore, the large regions of younger, smooth plains resemble the lunar maria in relative age, morphology, and mode of occurrence. Finally, both bodies are covered with a regolith of fragmental debris, which is not surprising considering that both are airless bodies battered over the eons by impacting objects of all sizes.

There the similarities end. The uplands of Mercury contain widespread regions of ancient intercrater plains—the planet's principal terrain. Although minor patches of old plains occur in the lunar highlands, they do not cover anything like the area covered by Mercury's intercrater plains. The widespread—probably global—system of lobate scarps is unique to Mercury. No other planet or satellite displays such a system of fractures. Albedos of various terrains on Mercury are systematically higher than their lunar counterparts, and the color characteristics are also different. The differences suggest that the Mercurian crust is relatively depleted in iron and titanium, compared to the lunar crust. Finally, certain characteristics of Mercurian craters—particularly their ejecta deposits—differ from those of lunar craters. All these differences are probably direct or indirect consequences of Mercury's huge iron core.

Chapter 6

A Battered World

Craters are the dominant landform in the Solar System. They occur to one extent or another on almost every solid body explored to date. Galileo in 1609 was the first person to recognize craters on the Moon. From that time until lunar exploration by spacecraft in the mid-1960s, a controversy raged over the origin of these landforms. One school of thought considered craters to be volcanic and the other believed they were the result of impacts. Over this 300-year period, little was known about volcanism and the mechanism of volcanic crater formation, and almost nothing was understood about impact crater formation.

In 1665, Robert Hooke dropped round bullets into a viscous mixture of clay and water and noticed that they formed craters like those on the Moon. He could not, however, understand where objects could come from to impact the Moon; at that time, meteorites were not known to come from space. Hooke heated dry gypsum powder in a pot. This released water vapor that rose in bubbles, leaving behind replicas of lunar craters. He therefore assumed that the lunar craters were due to some type of volcanism. G. K. Gilbert in 1892 was the first person to expound, in a scientific manner, the impact theory for lunar crater formation.

The controversy over lunar crater origin continued into the early 1960s and became so extreme that certain scientists in each camp insisted that either all lunar features were due to impact or all were due to volcanism. The reason this issue could not be decided to everyone's satisfaction was that the impact process was still not well understood, and the only photographs of the Moon were from Earth-based telescopes that were unable to resolve enough detail to prove an impact or volcanic origin. In the mid-1960s, the Ranger and the Lunar Orbiter missions obtained the first high-resolution images of the surface. At the same time, Donald Gault and his colleagues at the NASA Ames Research Center conducted cratering experiments using a high-velocity gas gun. From analyses of slow-motion films and study of the resulting impact craters, they derived the general sequence of cratering and the mechanics of impact crater formation. This new knowledge of impact cratering, combined with the acquisition of high-resolution images, provided the data necessary to answer the age-old question of lunar crater origin. It now became apparent to most scientists that the vast majority of lunar craters were the result of hypervelocity impacts. With more experiments, theoretical studies, and planetary missions, the impact origin for most Solar System craters has become well established. Although volcanic craters also occur on planetary surfaces, the criteria for distinguishing impact from volcanic craters are now well known.

Fresh impact craters are circular, with a floor that is considerably lower than the surrounding surface. Smaller impact craters are bowl-shaped, but larger ones have terraced inner walls and central peaks. The rim structure consists of a flap of overturned material, which is termed inverted stratigraphy. Impact craters are surrounded by an extensive ejecta deposit consisting of two parts: a relatively narrow inner zone of hummocky, continuous deposits, and an outer zone consisting of strings and clusters of secondary craters and rays. These secondary craters result from the impact of fragments or clots of material ejected from the crater. Ground observations usually show evidence of shock metamorphism in the form of high-pressure minerals, impact melt, planar features cutting across minerals, and peculiar cone-shaped structures

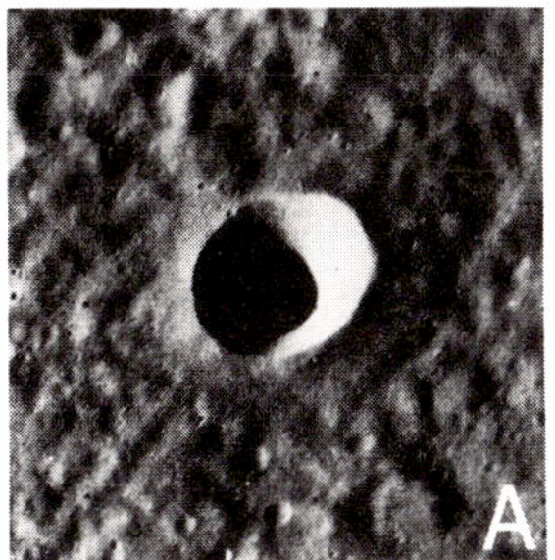

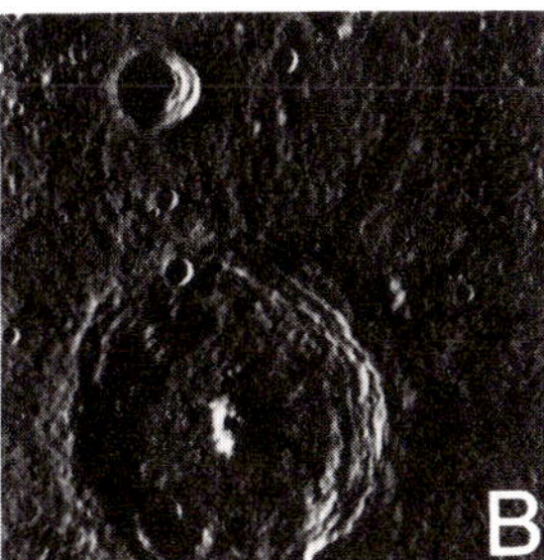

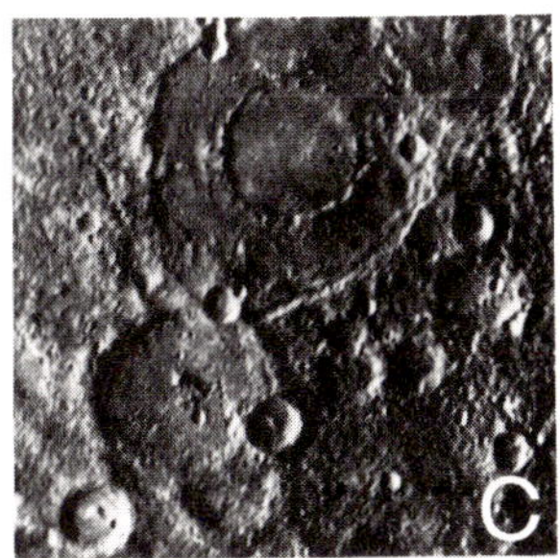

Crater morphology changes with diameter. At left (A) lies a simple bowl-shaped crater 8 kilometers in diameter. At diameters between about 15 and 100 kilometers, craters feature central peaks and terraces on the interior rims. The crater Brahms in the center (B) is such a complex crater, 75 kilometers in diameter. At diameters greater than 100 kilometers, craters develop central rings as shown by the 225-kilometer Bach basin at right (C). (Courtesy NASA.)

called shatter cones. In some cases, meteorite fragments are found around fresh craters such as Meteor Crater in Arizona.

Impact craters dominate Mercury's landscape. They are found on all types of terrain and in various states of preservation. The largest crater imaged by Mariner 10 is the Caloris basin, about 1,300 kilometers in diameter. Although Mariner 10 was unable to image features smaller than 100 meters, craters even less than a millimeter in diameter probably have formed by the impact of dust-size grains called micrometeorites, judging from the lunar sample evidence. More recent fresh-appearing craters feature extensive systems of bright rays that extend in some cases for thousands of kilometers across the surface. Other craters are so battered that they can be recognized only by the low, discontinuous remnants of their rims.

Crater Characteristics

Mercurian craters have a morphology—form and structure—similar in most respects to lunar craters. Their morphology changes with increasing diameter. At diameters less than about 20 kilometers, craters are bowl-shaped, with a depth that is about one-fifth their diameter. Craters between 20 and 90 kilometers in diameter are characterized by terraces on their interior walls and flat floors surrounding one or more central peaks. At diameters larger than about 90 kilometers, the floor structure begins to change from a single central peak to an irregular ring of peaks, and finally to a single concentric mountain ring.

The terrains surrounding Mercury's craters are hummocky surfaces consisting of material excavated from the crater. The width of this deposit, called the continuous ejecta blanket, is about half that for lunar ejecta deposits. Within and beyond the continuous ejecta blanket lie strings and clusters of small craters formed by the impact of individual fragments or clots of material ejected from the crater. These craters are called secondary impact craters. For extremely large impacts, secondaries can attain diameters of up to about 20 kilometers.

Crater Formation

This characteristic morphology of Mercurian and other Solar System craters results from the impact of objects, such as meteors and comets, with a surface. The impact process is probably the most devastating phenomenon that can affect the surface of a planet or satellite. When a high-speed object strikes, it produces energy that can reach enormous proportions; this energy is kinetic because it is caused by motion.

The amount of energy produced in this way depends on the mass of the impacting object and the velocity with which it strikes the surface. For instance, an iron meteorite 1 kilometer in diameter hitting a surface at a velocity of 15 kilometers per second will release more than 4×10^{27} ergs of energy—the equivalent of about 100,000 megatons of TNT (1 megaton = 1 million tons). This amount of energy is the same as that released by simultaneously exploding 20,000 5-megaton hydrogen bombs. The crater formed by such an event would measure about 10 kilometers in diameter, a small crater by Mercurian standards. Even some secondary impact craters are larger.

The explosion caused by an impact differs from that caused by a chemical detonation. In a chemical explosion it is the rapidly expanding gases that cause most of the damage. In an impact event the motion of the projectile (meteorite or comet) rapidly transfers kinetic energy to the target material (planetary crust). Most of this energy takes the form of shock or pressure waves that travel at supersonic speeds through both the target and projectile. These shock waves spread outward in a hemispherically expanding shell from the point of impact. The strength of the shock waves is so great that the rocks are subjected to pressures equivalent to those reached at depths of hundreds of kilometers inside the Earth. Pressures within the shocked rocks can rise to hundreds of kilobars (1 kilobar = 1,000 atmospheres).

Rocks are drastically affected by these extremely high pressures. For instance, granite is crushed at about 250 kilobars, and it begins to melt (from heat caused by compression) at 450 to 500 kilobars. At pressures above 600 kilobars it will vaporize. Minerals subjected to these pressures are compressed to higher-density phases, just as they are deep within the Earth (see chapter 4). Coesite and stishovite, two high-density forms of quartz, are sometimes found in terrestrial impact craters.

Pressure alone will not create large craters. It is the interaction of the shock waves with the unconfined surface—called a free surface—that is responsible for excavating the crater. After passage of the shock wave, the compressed rock snaps back along the free surface. This produces a tensional wave—called a rarefaction wave—that decompresses and fractures the rock, setting it into motion along fracture planes. The net effect momentarily converts the rock into a fluidlike material that moves lat-

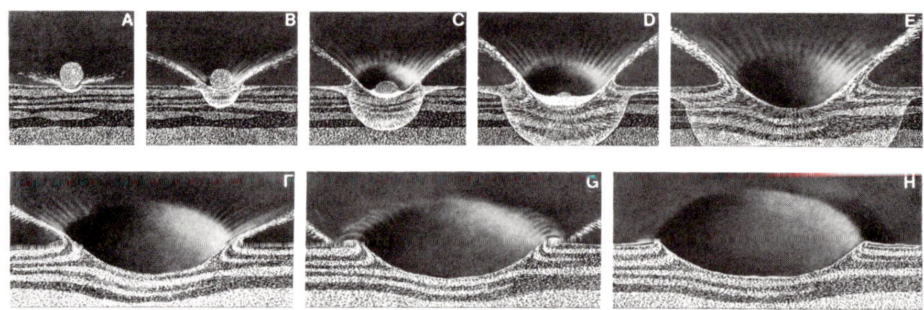

Stages in the formation of a simple bowl-shaped impact crater. In A, the initial contact causes highly shocked material to be jetted from the surface. In B, C, and D, a compressional shock wave moves outward in a spherical shell, and the crater grows by rarefaction behind the shock wave while the projectile is destroyed. In E, the crater continues to grow after the projectile is destroyed, and in F the crater reaches its maximum size. In G, the curtain of ejecta continues to expand outward after the crater stops growing, and an overturned flap of ejecta comes to rest on the crater rim. H represents the final configuration of the crater. (Drawn by Donald E. Davis.)

erally upward and out of a steadily growing excavation cavity. Thus, solid rock is fragmented and ejected from the growing crater in an expanding cone of rapidly moving material that is mostly deposited beyond the crater's final rim. The crater stops growing at a distance where the shock waves fall, or decay, below a critical value less than the strength of the rocks. This crater, called the excavation crater, can be enlarged by slumping of the rim. The rock layers at the edge of the crater are pushed upward and overturned by the passage of the rarefaction wave. This process and the deposition of ejecta result in the characteristic raised rims of impact craters.

Before this time, the projectile has suffered a similar fate. Shock waves generated in the projectile travel backward through its body and become transformed into rarefaction waves as they interact with the projectile's free surface. At these high pressures, the projectile essentially "explodes," some of it vaporizing, some melting into drop-

Meteor Crater in northern Arizona is one of the best-preserved impact craters on Earth. It measures about 1.3 kilometers in diameter and 180 meters deep. It was formed about 50,000 years ago.

lets, and the rest shattering into small pieces. Thus, crater-forming meteorites are preserved only as relatively small fragments widely scattered within and outside the crater. This is the reason the search was futile for a large meteorite thought to be buried below the floor of Arizona's Meteor Crater. In the early part of the century (1902–22) D. M. Barringer, a prominent industrialist, believed the occurrence of iron meteorite fragments around Meteor Crater indicated that a large, crater-forming, iron-nickel meteorite could be found mostly intact beneath the floor. He thought the meteorite would be large enough to be an economically important ore body, and he planned on reaping a large profit from mining it. An extensive drilling program on the crater's floor proved fruitless, and eventually the project was abandoned. Relics of the drilling equipment can still be seen on the floor of Meteor Crater. If Barringer had known then what we now understand about the physics of high-velocity impacts, he could have saved $500,000 and two decades of effort, as well as avoiding the wrath of his investors.

In large impacts, other effects can occur beyond the formation of an excavation cavity. The heat generated by a large impact can be so great that large volumes of the target material are melted. This melt is found as a sheet of once-molten rock overlying fragmented floor material (called a breccia) and as ponds and flows beyond the crater rim, deposited as part of the ejecta. The flat floors of fresh impact craters consist largely of impact melt. If the crater is large enough (greater than about 20 kilometers for Mercury), the wall of the excavation cavity becomes unstable and large parts of it slide inward, forming wall terraces and enlarging the original excavation crater. Central peaks also form in large craters, but the mechanism is not well understood. One possibility is that the sudden excavation of a crater causes the rocks beneath the impact point to undergo a shift from high pressure to low pressures in an extremely short period. This may cause the center of the floor to spring back or rebound into a central peak.

The size of an excavation crater depends not only on the amount of energy released by the impact, but also on the gravity field and certain physical properties of the projectile and target. For instance, on the same planet two impacts of equal energy in targets of different strength will produce craters of different size. A larger crater will be produced in a weaker material such as a regolith because the material is easier to break up. A larger crater will also form on a planet with a weaker gravity field because it is easier to excavate the material.

In all cases, a crater is many times larger than the projectile that formed it. Although the diameter of a crater depends on the complex interaction of many factors, a rough approximation is that the excavation crater will be about ten times larger than the projectile that formed it. We have seen, however, that in a large crater the excavation cavity is enlarged by inward slumping of the rim. The amount of enlargement depends on the size of the crater. With small craters of less than about 15 kilometers, little slumping occurs and the final crater is essentially the crater of excavation. With craters between 15 and 100 kilometers diameter, rim slumping can enlarge the excavation crater considerably. With truly large basin-forming impacts, entire sections of the crust collapse inward, enlarging the initial crater by several hundred kilometers.

Ejecta Deposits

The deposition of excavated material called ejecta takes two forms: a continuous covering of material from the crater rim outward to a distance of about one-half to one crater diameter (the continuous ejecta deposit), and swarms of secondary craters formed by the impact of individual or clots of fragments mostly beyond the continuous ejecta deposit. The area covered by continuous ejecta can be four to nine times the area of the crater. Individual fragments can be hurled for hundreds or thousands of kilometers. The bright ray systems associated with fresh craters consist of hundreds of small secondary craters with bright halos that together make up the rays. Powdery material created by the impact also contributes to the rays. If fragments are ejected at speeds exceeding the escape velocity of the planet or satellite, they will not return to the surface. Thus, crater ejecta can have far-reaching effects on the surface of a planet.

Individual particles of the ejecta travel on looping paths called ballistic trajectories, identical to the paths taken by artillery shells. For airless bodies, the distance that ejecta can be thrown on a ballistic trajectory depends both on the angle and velocity at which the material is ejected and on the gravity field of the planet. High-velocity impact experiments show that most ejecta is thrown from a crater

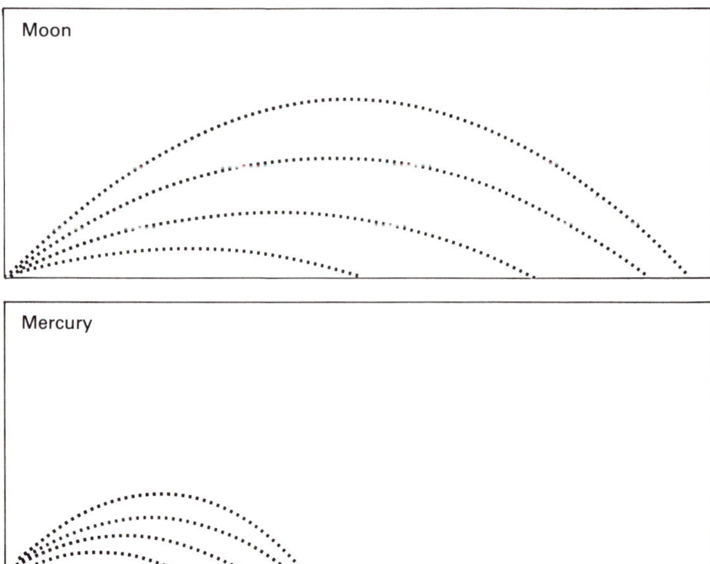

Because Mercury's gravity is stronger than the Moon's, impact ejecta travel more than twice as far on the Moon as on Mercury for similar-sized impacts.

at angles between 30 and 50 degrees from the horizontal. For a given ejection velocity, a fragment will travel farther when ejected at an angle of 45 degrees than one ejected at any other angle. (This explains why artillery must be raised to an angle of 45 degrees to attain the maximum range for a shell. At any other angle the shell will fall a shorter distance. So if the range to a target and the muzzle velocity of the shell are known, a gun can be pointed at the appropriate angle to hit the target.)

Gravity also governs the distance ejecta can travel. A fragment will travel farther on a planet with a weaker gravity field than one with a stronger field. This relationship accounts for one of the principal differences between lunar and Mercurian craters. Mercury's gravity field is almost twice as strong as the Moon's, so that for similar-sized impacts ejecta travel only half as far on Mercury.

The ejecta deposit of the lunar crater Copernicus (A) is more widely distributed than that of the similar-sized Mercurian crater (B) because of Mercury's stronger gravity field. (Courtesy NASA.)

Thus, on Mercury both the continuous ejecta deposit and a greater abundance of secondary craters are concentrated nearer the crater rim. Large secondaries can occur on the continuous ejecta deposit right up to the rim, whereas on the Moon they are concentrated beyond the continuous ejecta at a distance of about one crater diameter.

Crater Modification

The formation of craters and their central peaks, terraces, and ejecta deposits is part of an impact process that occurs within a time scale of seconds to minutes, depending on the size of the impact. Over time, however, craters and basins can be modified by processes that change their appearance. These processes can be both external or internal to the planet. On airless bodies such as Mercury, the modifications take three forms: subsequent impacts of both small and large objects, volcanic deposition, and crustal deformation. On planets with atmospheres such as Mars and Earth, wind and water erosion and deposition are also important modifying agents.

This photograph shows a group of similar-sized craters in various states of preservation. Crater 1 is the freshest, with a sharp rim and a pronounced ejecta deposit of secondary craters. Crater 2 has been degraded by subsequent cratering and the ejecta deposit of Crater 1. Crater 3 is even more degraded by subsequent impacts, secondary cratering, and the flooding of its southern rim by intercrater plains. (Courtesy NASA.)

Impacts degrade preexisting craters by chipping away at their structures and depositing ejecta on them. Large impacts destroy all or large parts of preexisting craters, and if the process operates long enough, they degrade even large craters to the point where they are barely discernible.

Volcanic activity also modifies the appearance of craters. Extensive flooding by lava can cover ejecta deposits and fill a crater's interior so that only a low rim protrudes through the lava flows. If the lava flows are extensive enough and extruded over long periods, they can completely bury even large craters.

Crustal deformation changes a crater by shifting the rim or floor and distorting its shape. All these modifying

Computer-generated photomosaic of Mercury's Caloris basin (Courtesy NASA.)

processes produce craters in various states of preservation: from relatively recent, fresh-appearing structures with sharp rims and extensive ray systems to degraded craters with barely disernible rims.

The Caloris Basin

Basin-forming impacts are the most devastating events that a planetary surface can experience. Their effects are so widespread that few areas of a planet are left unscathed. They can also trigger internal processes that affect large areas of a planet. The formation of the Caloris basin was such an event in the history of Mercury. This basin constitutes one of the largest impact craters yet discovered in the Solar System. It measures 1,300 kilometers in diameter, but it has affected a wide area well beyond the basin rim. Even a large region on the opposite hemisphere has been severely disturbed by the original impact.

The Caloris basin consists of a ring of mountains rising about 2 kilometers above its floor. Another faint cliff or scarp occurs about 150 kilometers beyond the main mountain ring northeast of the basin. This scarp is probably a fault along which a segment of the crust slid inward toward the basin's center just after the impact. In this sense, it resembles terraces formed in smaller craters. The area between the scarp and the main mountain ring consists of blocky terrain, probably caused by the breakup of the crust as it slid toward the basin. Beyond this faint scarp stretches a system of valleys, which radiate from the basin for about 1,000 kilometers. These valleys may be radial fractures or chains of coalescing secondary impact craters formed by clots of fragments ejected from the basin. Numerous crater clusters, chains, and irregular troughs lie within the cratered uplands beyond a distance of one basin diameter. Some of these attain diameters of 20 kilometers; they, too, are probably Caloris secondary craters.

Just outside the basin rim are several large regions of hummocky plains that extend outward several hundred

Detail of the Caloris basin's northeastern rim. The main ring is located at A-A[1] and a weaker outer ring can be detected at B-B[1]. (Courtesy NASA.)

Linear valleys and ridges radiate from the Caloris basin. They were probably formed by ejecta from the basin. (Courtesy NASA.)

These rolling plains with interbedded hills are thought to be a continuous ejecta deposit from the Caloris impact. (Courtesy NASA.)

kilometers. These regions include numerous hills and therefore differ from other Mercurian plains. The hills are probably remnants of the continuous ejecta deposit, composed of a mixture of large fragments and partially melted ejecta. Surrounding these ejecta deposits are smooth plains that occur up to 2,500 kilometers from the basin. They occupy low-lying areas and may be volcanic deposits similar to the lunar maria that also fill and surround impact basins.

The floor of the Caloris basin is filled with smooth plains of a structural pattern and intensity unlike any other basin in the Solar System. Numerous ridges form a pattern that is both concentric and radial to the center of the basin. These are similar to ridges found in the lunar maria, but

The Caloris basin floor contains a complex pattern of ridges and fractures. (Courtesy NASA.)

in Caloris they are much more numerous and feature a strong radial component not observed on the Moon. As on the Moon, however, these ridges were probably caused by compressive forces. The ridges are cut by a younger system of tensional fractures called graben, which also form a concentric and radial pattern. The fractures measure up to 10 kilometers wide and progressively increase in width and depth toward the basin's center.

These two very different structures were probably caused by vertical movements of the floor following impact. The ridges formed first, as the floor subsided along inward sloping fault planes at the margin of the basin. Since the floor covers such a large area of the planet (30 degrees of latitude), it has a substantial outward curvature.

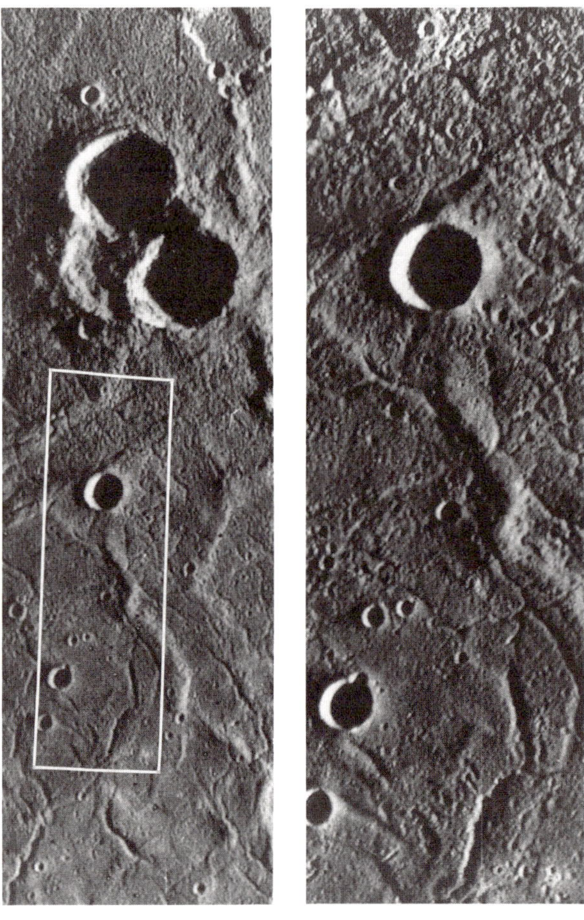

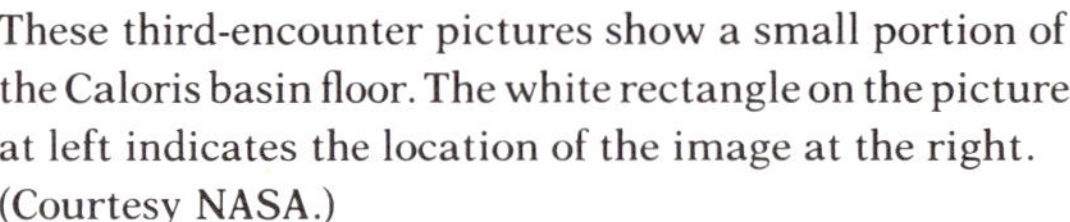

These third-encounter pictures show a small portion of the Caloris basin floor. The white rectangle on the picture at left indicates the location of the image at the right. (Courtesy NASA.)

This strip of two third-encounter images shows the complexity of the ridge and fracture systems that cover most of the Caloris basin floor. (Courtesy NASA.)

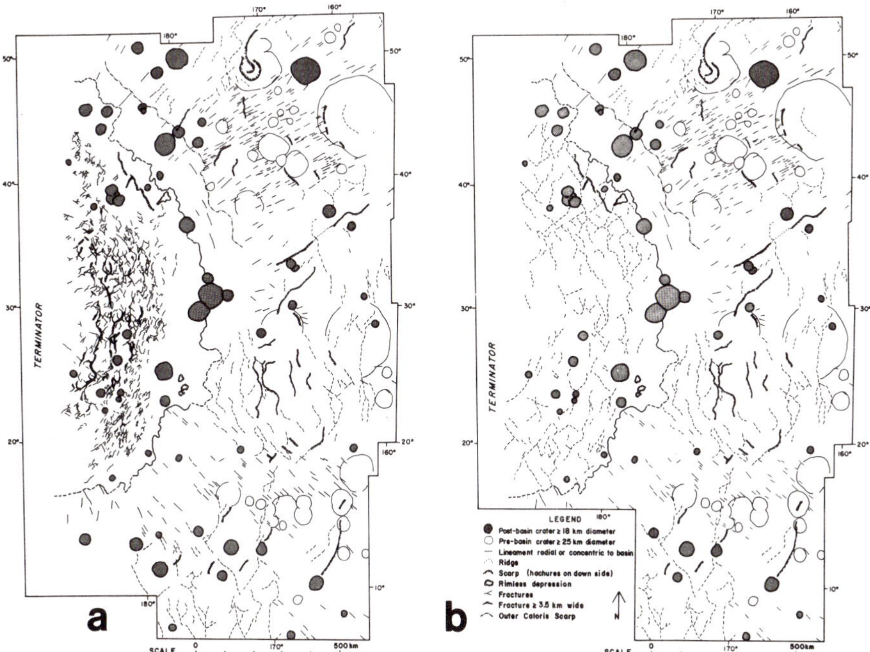

These maps show the pattern of fractures (a) and ridges (b) on the Caloris basin floor. (From Strom, Trask, and Guest, *Journal of Geophysical Research*, Vol. 80: 2,478, 1975.)

As the floor subsided, it was squeezed into a smaller area, causing compressive stresses that wrinkled the surface with ridges. Subsequently, the floor was uplifted, stretching the surface and causing tensional stress that produced the fractures. These fractures are the only evidence of tension on the planet and therefore make Mercury unique among the bodies in the inner Solar System. The cause of these vertical movements is not known. They may have resulted from lateral and vertical migrations of subsurface magmas (molten rock), adjustments of the floor due to gravity, the weight of heavy lavas, or some other cause.

The Hilly and Lineated Terrain

On the other hemisphere of Mercury, directly opposite the center of the Caloris basin, lies a peculiar, broken-up

The region of Mercury's hilly and lineated terrain lies at the antipodal point of the Caloris basin. The outlined area can be seen in more detail in the next illustration. (Courtesy NASA.)

terrain covering an area of at least 360,000 square kilometers. It extends beyond the terminator into the unlit hemisphere and may be twice as large. It consists of hills, depressions, and valleys that have partially destroyed preexisting landforms. The hills are 5 to 10 kilometers wide and up to 2 kilometers high. Some valleys are more than 120 kilometers long and 15 kilometers wide. They generally trend at right angles to each other in northwest and northeast directions. Many craters' rims and floors have been broken up into hills and depressions. In some

Detail of the hilly and lineated terrain. The outlined area can be seen at high resolution in the next illustration. (Courtesy NASA.)

cases, disrupted craters' floors are filled with younger, smooth plains.

This terrain almost surely was caused by the impact that formed the Caloris basin. An impact the size of Caloris generates tremendous seismic ("earthquake") energy, in the form of waves that travel through the planet and along its surface. These seismic waves converge or focus at the antipodal region—the area directly opposite the impact on the other side of a planet. Computer simulations of seismic wave propagation for impacts of this size show that the seismic effects in the antipodal region can be enormous. The effects are even more pronounced when the waves travel through a large, high-density core such as Mercury's. The ground experiences vertical movements greater than 1 kilometer in a matter of minutes, and tension fractures tear the crust to depths of tens of kilometers.

High-resolution view of the hilly and lineated terrain. (Courtesy NASA.)

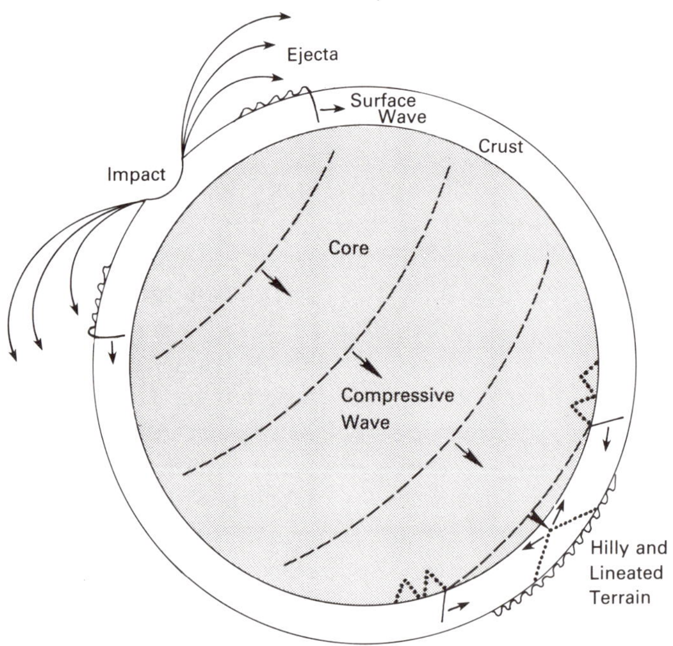

Seismic waves generated by the Caloris impact were focused at the antipodal point, causing large vertical ground movements and the hilly and lineated terrain. (Courtesy Peter Schultz.)

This stress would break the surface into a jumble of blocks and depressions as seen in Mercury's hilly and lineated terrain. Thus, the Caloris impact had far-ranging effects that scarred almost an entire hemisphere and disrupted a large region of the opposite hemisphere.

Catastrophe

For a person living on our rather tranquil Earth, it is difficult to imagine the catastrophic effects that would result from a Caloris-sized impact. Such events seem far removed in both space and time and lie completely out of the realm of human experience. But such impacts did occur on Earth early in its history. We do not really know all the ramifications of such a large terrestrial impact, but based on our observations of other planets we have evolved at least a general idea of what would happen.

Let's suppose that a planetesimal strikes the Earth near the center of Texas with such force that it creates a crater

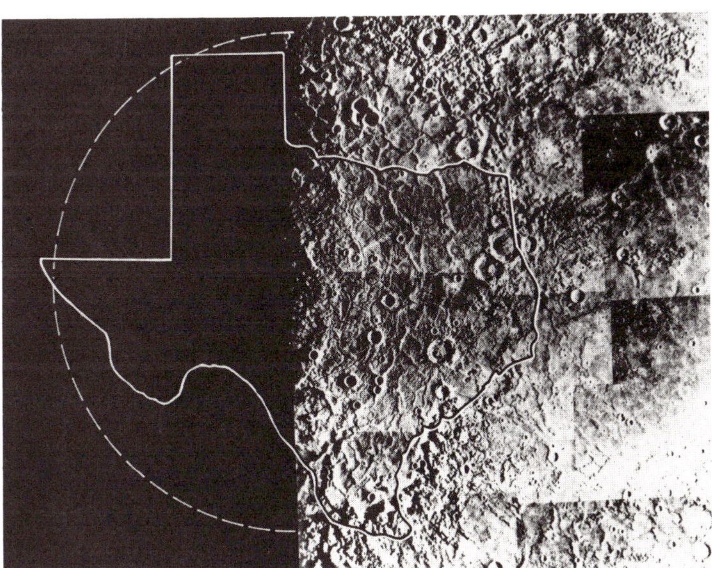

The Caloris basin would engulf the entire state of Texas. (Courtesy NASA/Goddard Space Flight Center.)

the size of the Caloris basin. The crater formed by the impact would obliterate the entire state of Texas and portions of adjoining states and Mexico. A continuous blanket of partially molten ejecta would cover New Mexico, Colorado, Kansas, Oklahoma, Arkansas, Louisiana, Mississippi, and a large part of Mexico, with thicknesses of debris ranging from a few hundred meters to more than 2 kilometers. Clots of ejecta would be thrown as far as Chicago and California, some forming craters large enough to engulf entire cities. Smaller secondaries ranging from hundreds of meters to a kilometer in diameter would pepper a large part of the United States and Mexico. A tsunami or tidal wave generated in the Gulf of Mexico would inundate Florida, the West Indies, the east coast of Mexico and Central America, and the coasts of Colombia and Venezuela.

But this is just the beginning. Seismic waves would cause earthquakes that would destroy most cities and towns in North America. They would also cause large vertical ground motions and disruption in the antipodal region near the center of the Indian Ocean. This would generate another tsunami that would inundate the coasts of western Australia, Indonesia, India, and Africa. Fractures beneath and surrounding the basin would penetrate the Earth's asthenosphere to produce volcanic flooding in the basin and probably in adjacent areas as well. An air shock wave produced by the passage of the planetesimal through the atmosphere would raise the air temperature for a short time to levels that would largely vaporize the exposed biosphere.

Following all this, the Earth's atmosphere would be so filled with dust from the impact that the Sun's rays would be largely blocked, and temperatures would fall drastically to initiate an ice age. Mass starvation and death would decimate what life was able to survive the initial impact. These are probably only some of the effects that would happen to Earth. Others might have even more far-reaching consequences, such as altering the movements of Earth's tectonic plates.

The likelihood of such a large impact occurring today is extremely remote. However, small to medium-sized impacts have occurred in geologically recent and even historical times. Strong evidence now exists that an asteroid or comet about 10 kilometers in diameter struck the Earth 65 million years ago and may have been responsible for the mass extinction of many species, including the dinosaurs. Only 49,000 years ago an impact formed Meteor Crater (1.2 kilometers in diameter) in northern Arizona, and on June 30, 1908, a small comet was apparently responsible for the devastation of a 10-kilometer area in the unpopulated region of Tunguska in Siberia. Currently, there is a program to search for Earth-crossing asteroids to more accurately determine their number, size, and chances of hitting Earth.

Origin of the Impacting Objects

We do not yet know for certain the origin of the objects that were responsible for the cratering record in the Solar System. Comets and some asteroids are presently the only small objects that cross planetary orbits. They have surely contributed to the cratering record. But are they the main source? Recent studies of the Solar System cratering record and the dating of returned lunar rocks have shed some light on this question.

The manned Apollo expeditions to the Moon returned rocks from a variety of locations. These specimens were used, among other things, to determine the absolute ages of the lunar surface. From these samples it was learned that the sparsely cratered mare lavas are ancient, dating from 3.9 to 3.0 billion years old. The heavily cratered highlands are even older, dating from 4.4 to 4.0 billion years old. The lunar highlands therefore accumulated their great abundance of craters, including the large mare-filled basins, over a geologically short span of 400 million years. On the other hand, the younger lunar maria accumulated their much smaller number of craters over the enormous span of 3 to 4 billion years, or about ten times longer. These data suggest that early in its history, the Moon underwent a period of intense bombardment that ended about 4 billion years ago. It was during this early period of heavy bombardment that the large basin-forming impacts occurred. Since about 4 billion years ago, large impacts have been infrequent and no large basin-forming events have occurred.

Early Mariner and later Viking missions showed that the southern hemisphere of Mars is largely a heavily cratered surface similar to the lunar highlands. The northern hemisphere is predominantly young plains, with a low crater abundance similar to the lunar maria. These observations suggest that, like the Moon, Mars also experienced the same heavy bombardment early in its history, followed by a sharp decline in the impact rate.

At the time of these observations, it was widely believed that the source of the impacting objects was the asteroid belt situated between the orbits of Mars and Jupiter. Meteorites, which probably come from the asteroid belt, frequently hit the Earth and must also strike the Moon and Mars. We know that today some asteroids still cross the orbits of Earth and Mars. It seemed logical that the heavily cratered regions of the Moon and Mars could have been formed by asteroids, if the asteroids had been colliding with each other in the belt more frequently in the past, sending the debris into the inner Solar System. It was reasoned that as mutual asteroid collisions became less frequent with time, fewer fragments would be sent to the inner planets, accounting for the lesser abundance of craters on the younger Martian plains and lunar maria. Asteroids collide with Mercury much less frequently than with Mars and the Moon, because it is much more distant from the asteroid belt. If the craters on the Moon and Mars were caused mainly by asteroids, then Mercury should have a crater abundance about half that on Mars.

To the surprise of many scientists, Mariner 10 revealed a surface just as heavily cratered as that of the Moon and

Mars. This discovery had profound implications for the cratering history of the inner planets. It cast doubt on the idea that heavily cratered terrains were caused principally by asteroids. Furthermore, since Mercury, the Moon, and Mars span the entire range of inner planet distances (0.38 to 1.5 AU), there was no doubt that all the terrestrial planets had experienced a period of heavy bombardment early in their histories. Earth and Venus could not have escaped this devastating bombardment.

If Earth was so heavily impacted, then where are all the craters that should have formed during this period? Earth is an extremely dynamic planet whose surface is constantly changing. It is intensely eroded by weathering processes. The crust is constantly being deformed, and in some places completely destroyed, on a geologically short time scale by a process called plate tectonics. About 60 percent of the Earth's surface (the ocean basins) has been created by a plate tectonic phenomenon called seafloor spreading during the past 250 million years—only the last 5 percent of geologic history. Even the oldest areas on Earth—the continental Precambrian shields—are only 1 to 3 billion years old. But the period of heavy bombardment ended about 4 billion years ago. Therefore, the craters formed during this period have been completely erased from Earth.

On Venus, however, the Pioneer Venus radar data and Earth-based radar results show some areas with large circular depressions that may be impact craters. Radar images of the planet's northern hemisphere from the Soviet Venera spacecraft reveal that impact craters are present on Venus, but their number is considerably less than on the Moon, Mars, or Mercury.

When the two Voyager spacecraft flew by Jupiter, Saturn, and Uranus, they discovered that the satellites of these planets also have heavily cratered surfaces. This suggested that probably the entire Solar System was subjected to a period of heavy bombardment early in its history, and that the asteriods were not the major cause of

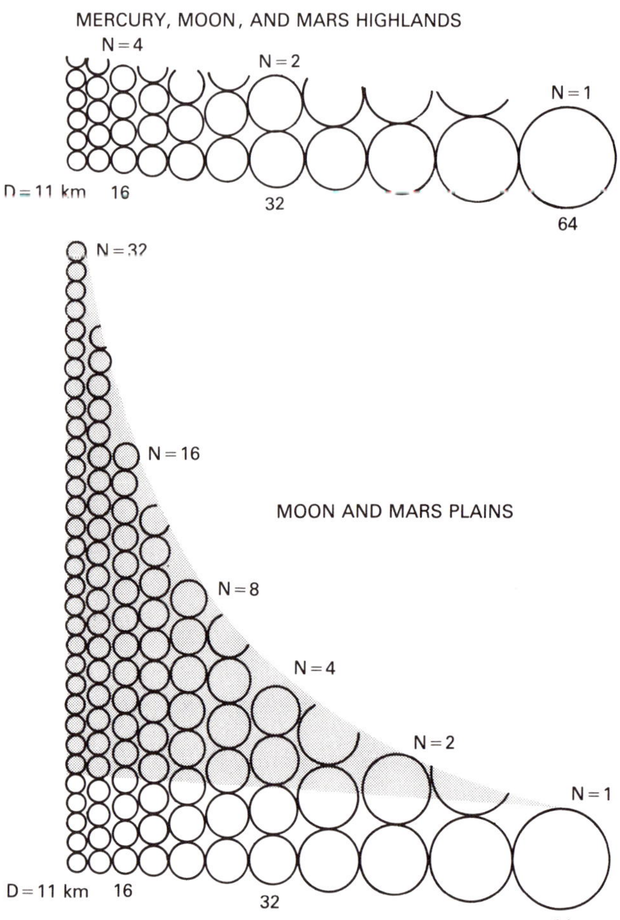

This diagram illustrates the difference between the size/frequency distributions of craters between 11 and 64 kilometers in diameter found on the terrestrial planets. The size distribution in the upper diagram represents the heavily cratered highlands that resulted from the period of heavy bombardment. For every crater 64 kilometers in diameter there are two at 32 kilometers and seven at 16 kilometers. The size distribution in the lower diagram represents the lightly cratered plains on the Moon and Mars. Here, for each crater 64 kilometers in diameter there are four at 32 kilometers and sixteen at 16 kilometers. The shaded area indicates the difference between the two crater populations. The highlands crater population may have resulted from remnants of the objects from which the planets accreted, while the plains crater population is probably the result of comet and asteroid impacts.

the cratering record. Asteroids are fairly easy to perturb into the inner Solar System because the Sun's strong gravity tends to pull them inward. But they are quite difficult to perturb into the outer Solar System because they would then be fighting the Sun's large gravity. Only a few of the craters on Jupiter's satellites can be due to asteroids and much less, if any, on Saturn's and Uranus's satellites.

If the asteroid belt is not the main contributor to the heavily cratered terrains, then where could the impacting bodies come from? Although we are far from sure, some clues to their origin can be deduced from the size/frequency distribution of the craters. Size/frequency distribution measures the number of craters of a given diameter per unit of surface area. One simply calculates the diameter of the craters in an area and determines the number that lie within various diameter intervals. The number of craters per unit area—called the crater density—also provides a measure of the age of the surface; the greater the crater density, the older the surface.

All the heavily cratered surfaces on the Moon, Mars, and Mercury have essentially the same size/frequency distribution. This relationship indicates that the same family of objects was responsible for the period of heavy bombardment in the inner Solar System. On the other hand, the younger lunar maria and Martian plains have a different crater size/frequency distribution, characterized by an overabundance of small craters compared to the heavily cratered surfaces. This pattern suggests that a second family of objects was responsible for the cratering on surfaces younger than 4 billion years.

Thus, two families of objects appear to have caused the cratering in the inner Solar System: one responsible for the early period of heavy bombardment, and another group impacting much less frequently from about 4 billion years ago up to the present. This later group is probably a combination of asteriods and comets. The earlier family must have consisted of many large objects that formed the impact basins, yet were swept up and became extinct in a geologically short period—about 400 million years. One suggestion is that these objects were comets that were much more abundant early in Solar System history. An even more intriguing possibility is that they were remnants from the formation of the terrestrial planets themselves. Computer studies suggest that material accreting into the inner planets would be swept up on a short time scale (about 70 million years), but that a portion of this material would last for about the observed time span of 400 million years.

The cratering record in the outer Solar System is more complex. It is characterized by a dearth of large craters compared to the terrestrial planets, and different crater size/frequency distribution on the satellites of Jupiter, Saturn, and Uranus. The small number of large craters suggests that the objects responsible for the period of heavy bombardment in the outer Solar System were different from those that caused the heavy bombardment in the inner Solar System. If all the heavily cratered surfaces were caused by comets, then it is difficult to explain the different crater size/frequency distributions in radically different regions of the Solar System. Possibly the heavily cratered surfaces in the outer Solar System were caused by local impacting objects associated with Jupiter, Saturn, and Uranus.

Much more research must be carried out before we can determine the origin of the objects responsible for cratering the Solar System.

Chapter 7

Plains: Smoothing the Rough Spots

Plains constitute a common type of terrain on almost all solid bodies in the Solar System. On Mercury they occupy a larger proportion of the surface (about 60 percent) than any other terrain type. The Mercurian plains have been classified into two categories: smooth and intercrater. This division is largely based on differences in the abundance of superposed craters. Intercrater plains are much more heavily cratered than smooth plains and, therefore, are older. Mercury's plains differ from lunar plains in two important respects: they are much more widely distributed, and they have a significantly higher albedo. Plains can form by various processes such as erosion and deposition, volcanism, and impact ejecta mantling. The origin of Mercury's plains remains somewhat uncertain, but the current evidence seems to favor a volcanic origin.

Smooth Plains

In some respects smooth plains resemble the lunar maria. They are relatively featureless, rolling surfaces with a low abundance of impact craters. The larger expanses of smooth plains are often crossed by ridges similar to the wrinkle ridges found on the lunar maria. Irregularly shaped, rimless pits from 5 to 10 kilometers in diameter occur in two areas of Mercury's smooth plains. These resemble pits in the lunar maria that have been attributed to volcanic processes. Smooth plains occupy about 15 to 20 percent of the territory viewed by Mariner 10, with the

Most of this area is covered by smooth plains that surround the Caloris basin. A ridge similar to lunar wrinkle ridges crosses the middle of the photograph. (Courtesy NASA.)

most extensive areas of smooth plains occurring in and around the Caloris basin and in the north polar region.

Earth-based radar profiles south of the Caloris basin show that the smooth plains, at least in this region, extend well beyond the terminator viewed by Mariner 10. These data suggest that the plains may completely surround the Caloris basin. Smaller patches are scattered among the heavily cratered uplands. All smooth plains are situated in low-lying areas. Earth-based radar altimetry indicates that the smooth plains outside the Caloris basin lie about 2.5 kilometers below the surrounding terrain. As on the Moon, the larger areas of smooth plains are mostly associated with impact basins. They fill and surround the Caloris basin and occur on the floors of smaller basins such as Tolstoj (400 kilometers) and Dostoevskij (390 kilo-

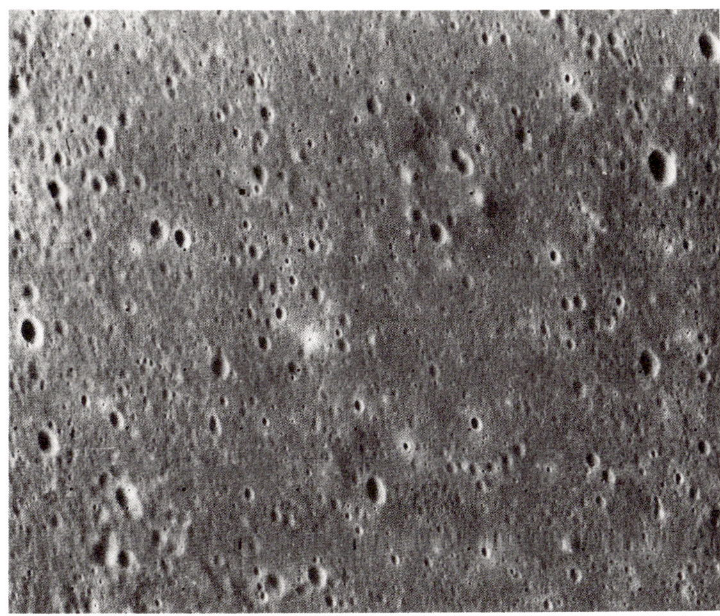

High-resolution photograph of the smooth plains. The smallest crater that can be seen measures about 200 meters across. (Courtesy NASA.)

meters). Even the north polar plains may fill a degraded basin 1,000 kilometers in diameter.

The age of smooth plains relative to other terrains can be determined from overlap relationships and the abundance of impact craters superposed on their surfaces. Smooth plains cover or embay preexisting topography and feature a lesser overall crater abundance than other Mercurian surfaces. This indicates that smooth plains are the youngest surfaces on the planet. However, they also have almost double the crater abundance of the lunar maria; if the impact rates on Mercury and the Moon were similar, then Mercury's smooth plains may be older than the lunar maria.

Mercury's smooth plains are also younger than the impact basins with which they are associated. The rims of large craters sometimes poke through the smooth plains on the basin floors, and the crater density on these plains

Smooth plains fill the center of this old impact basin named Tolstoj after the famous Russian writer. The arrows point to several craters that have been flooded by the smooth plains. The "A" indicates the location of an elongated, rimless depression that might be a volcanic crater. (Courtesy NASA.)

These smooth plains occur in the north polar region and appear to fill an ancient impact basin about 1,000 kilometers in diameter. They also fill a younger basin located in the upper left corner of this photomosaic. (Courtesy NASA.)

is substantially less than on the basin rim areas. This shows that the basins formed first and were subsequently impacted by smaller objects that cratered their floors and rims. After the impact rate had fallen, smooth plains were deposited on the basin floors and completely or partially buried preexisting craters. Crater abundances on the smooth plains surrounding the Caloris basin also indicate they are younger than the basin itself. Thus, like the lunar maria, Mercury's smooth plains are generally younger than the impact basins with which they are associated.

The albedo or brightness of the planet's smooth plains varies. Those surrounding the Caloris basin are darker than others, while some patches are as bright or brighter than the uplands in which they lie. These albedo differ-

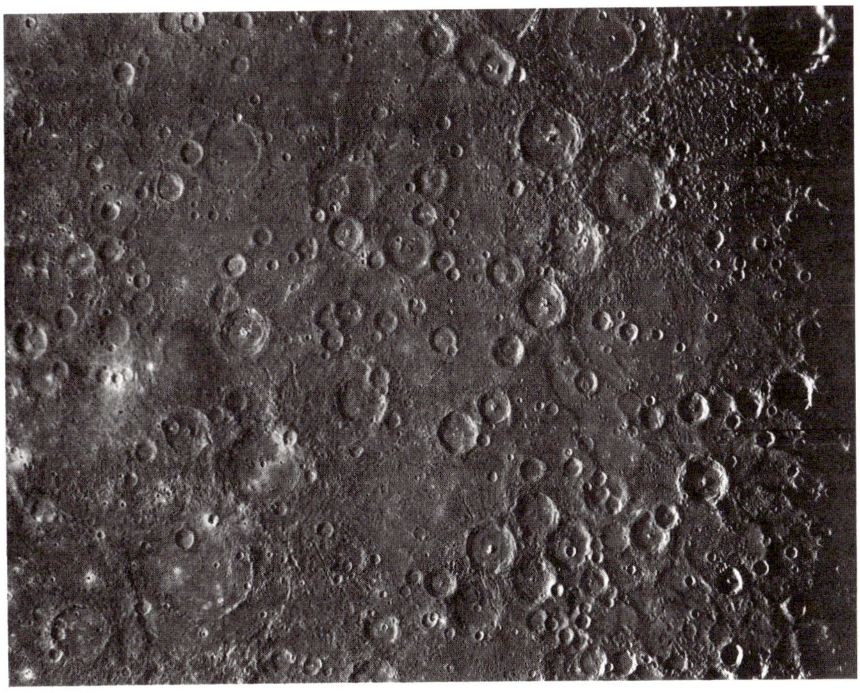

Intercrater plains that occur between and around regions of heavily cratered areas dominate most of Mercury's highlands. (Courtesy NASA.)

Large areas of intercrater plains (IP). (Courtesy NASA.)

ences probably result from differences in composition. All Mercurian smooth plains, however, are brighter than the lunar maria. This brightness, together with color differences, has been attributed to a lower abundance of titanium and iron in Mercurian rocks.

Intercrater Plains

The main differences between intercrater plains and smooth plains are their ages and distribution. Intercrater plains feature level to gently rolling surfaces with a large number of superposed craters less than about 15 kilometers in diameter. The large number of small superposed craters gives these plains a rougher texture than the smooth plains and attests to their greater age.

Intercrater plains occupy about 45 percent of Mercury's upland surface viewed by Mariner 10. They occur between

A high-resolution image of intercrater plains. These plains have flooded a large crater (70 kilometers) indicated by the arrow. (Courtesy NASA.)

and around clusters of large craters in the more heavily cratered terrain. Many of the superposed small craters form chains or clusters, suggesting they are secondary impact craters resulting from the larger craters that make up the heavily cratered terrain. The high density of superposed craters indicates that they predate the smooth plains and that they constitute one of the planet's oldest surfaces.

Intercrater plains apparently span a range of ages relative to the period of heavy bombardment. They partially bury some large craters and the ejecta blankets of others, but other large craters and their ejecta are superposed on the plains. This relationship indicates that intercrater plains formed early in Mercury's history during the period of heavy bombardment. The crater size/frequency distribution also suggests that the intercrater plains were deposited during heavy bombardment; there is a lesser abundance of craters smaller than 50 kilometers diameter on the uplands compared with the lunar highlands, revealing a systematic loss of smaller craters. This pattern is exactly what would be expected if Mercury's intercrater plains were formed during the heavy bombardment: smaller craters would be buried first, followed by larger craters as the plains continued to be deposited. Intercrater plains on the Moon are extremely rare, so this type of crater obliteration did not occur there to the same extent. Craters superposed on the smooth plains register the last phase of cratering on Mercury, and they do not show this type of loss.

Intercrater plains have been dated relative to craters of various ages. More intercrater plains are associated with the older craters than with the younger ones. Apparently the volume of these plains decreased as the impact rate declined. The brightness of intercrater plains is about 25 percent greater than that of the smooth plains surrounding the Caloris basin. They therefore may have a lower titanium and iron content than the smooth plains.

Origin of the Plains

Two completely different origins have been proposed for the Mercurian plains. One hypothesis considers the plains to be impact ejecta deposits from large basins. The other considers the plains to be primarily volcanic deposits.

Just beyond the continuous ejecta deposit of fresh lunar basins lie small regions of light, featureless plains. These deposits are probably fluidized ejecta that flowed beyond the continuous ejecta, partially burying preexisting craters. At greater distances, patches of light plains partially fill craters and other low areas in the lunar highlands. Apollo 16 returned samples from one of these areas that proved to be impact breccias, probably from an ejecta deposit of the Imbrium basin.

By analogy with the Moon, both the Mercurian smooth and intercrater plains have been interpreted by some scientists to be composed of basin ejecta. According to this hypothesis, the smooth plains surrounding the Caloris basin are smooth ejecta deposits from the basin. The smooth plains filling the basin would be either impact melt formed on the basin floor during the impact event, or ejecta deposited on the floor by other basin-forming impacts. The intercrater plains would have been created by the same type of ejecta deposits formed earlier in Mercury's history during the period of heavy bombardment.

There are, however, several problems with this hypothesis. First, the lunar light plains ejecta deposits cover only about 5 percent of the Moon's surface, but Mercury's smooth plains cover about 15 percent of the surface viewed by Mariner 10. It is difficult to explain why Mercury's ejecta deposits should be so much more extensive than the Moon's. The Caloris smooth plains extend outward in a more-or-less continuous spread, covering some areas more than 2,000 kilometers from the basin rim, while no such extensive ejecta deposits occur around lunar basins of comparable size. Since the ballistic range on Mercury is about half that on the Moon, the ejecta deposits should be concentrated closer to Mercurian basins—as is observed for the smaller craters. Also, large fresh Mercurian craters greater than 200 kilometers do not show proportionately large areas of smooth plains surrounding them. Instead, their ejecta deposits consist of narrow, continuous deposits with numerous clusters and strings of secondaries right up to the crater rims.

The observation most damaging to the ejecta hypothesis is the relative age of smooth plains and their associated basins. Ejecta deposition and impact melting are part of the impact process, and therefore ejecta and impact melt deposits should be exactly the same age as the crater or basins from which they originate. Large areas of Mercury's smooth plains, however, are younger than their associated basins and therefore cannot be ejecta or impact melt from the original impacts that formed these basins.

The Caloris smooth plains are also darker than other smooth or intercrater plains, which suggests a compositional difference. The albedo is more compatible with a volcanic origin than an impact ejecta origin. The distribution of Mercurian smooth plains is in fact quite similar to that of lunar mare volcanic deposits; both are associated with older impact basins and have about the same areal distribution. Earth-based radar data indicate that the annulus of the smooth plains surrounding the Caloris basin is strongly down-bowed, like the lunar maria. On the Moon, this is due to the weight of volcanic deposits. These combined observations favor a volcanic origin for the majority of Mercurian smooth plains.

Somewhat similar arguments can be made for a volcanic origin of the intercrater plains. They cover about 45 percent of the surface viewed by Mariner 10, but there are no evident source basins from which they could be derived. On the Moon, small patches of similar plains exist in the highlands, but they constitute less than 3 percent of the surface. The origin of these lunar deposits is still debatable, but unlike Mercury's, they can at least be identified with possible source basins.

Landforms of a clear-cut volcanic origin are rare on Mercury, but one must keep in mind that the Mariner 10 resolution and coverage of the planet is only about the same as Earth-based prespacecraft telescopic resolution and coverage of the Moon. At this resolution, there are few identifiable volcanic landforms on the Moon, although most of the lunar front side is covered by volcanic deposits. Nevertheless, several landforms on Mercury have been tentatively interpreted as volcanic. In the smooth plains on the floors of the Caloris and Tolstoj basins are several irregularly shaped, rimless pits similar to volcanic pits in the lunar maria.

The only way to prove whether or not Mercury's plains are volcanic is to examine returned samples from its surface. Until then, the evidence seems to favor a volcanic origin for most of Mercury's smooth and intercrater plains. If so, then Mercury experienced a period of volcanism much more extensive than the Moon's. This would be consistent, however, with thermal history models that have predicated much greater heating and melting of Mercury than the Moon.

Mode of Plains Formation

If Mercury's plains are volcanic, then how were they deposited and why do they cover such a large area of the planet? Most people tend to think of volcanism in the form of large cones with summit craters, such as snow-capped Mt. Fujiyama in Japan or Mt. Kilimanjaro in Africa. These are probably the most beautiful structures created by volcanic action, but they represent only one of a variety of landforms produced by volcanism.

The form that volcanism takes is highly dependent on the composition of the lava and the volume and rate at which it erupts. Lavas rich in silica, aluminum, sodium, and potassium and poor in magnesium and iron are extremely viscous and form short, thick flows or domes. They usually contain a large amount of gas that can be released explosively to form extensive ash deposits. Eruptions of this material form alternating layers of ash and lava that build large conical mountains called composite or stratovolcanoes. Mt. Fujiyama is an example. Lavas rich in iron and magnesium but poor in silica, aluminum, sodium, and potassium are quite fluid. This type of lava, called basalt, usually has a small gas content and therefore produces little ash. Basalt eruptions build broad domical mountains called shield volcanoes. Their summits usually end in a large collapse crater called a caldera. The best-known example of a shield volcano is Mauna Loa on the island of Hawaii.

Basaltic volcanism can also produce extensive plains called plateau or flood basalts. These plains are formed by enormous volumes of fluid basalt erupting from long fissures. Individual flows can cover areas more than 40,000 square kilometers. The fissures that gave rise to these flows are usually buried by the lava, and therefore the sources

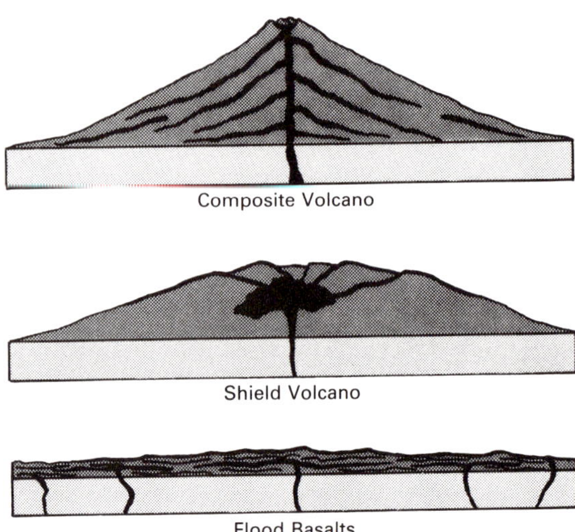

Three major volcanic processes form composite volcanoes, shield volcanoes, and volcanic plains (flood basalts). Composite volcanoes are made of layers of ash and lava, forming steep cones. Shield volcanoes are made only of fluid basaltic lavas that create gentle-sloped domes. Flood basalts are composed of voluminous fluid basaltic lavas that erupted from long fissures, forming extensive plains. Only volcanic plains are prevalent on Mercury.

of the flows are rarely visible on the surface. Large deposits of flood basalts occur in India, Brazil, Africa, and the United States. In the United States the Columbia River basin's flood basalts cover about 220,000 square kilometers of Washington, Oregon, and Idaho—deposits that were erupted about 15 million years ago. Individual flows averaged about 25 meters thick, and the estimated volume of these deposits totals 195,000 cubic kilometers.

The Moon's maria are flood basalts that were erupted over a period of more than one billion years. They cover about 17 percent of the lunar surface with an estimated volume of 10 million cubic kilometers. Some individual lava flows have traveled more than 350 kilometers. As with terrestrial flood basalts, only in a few cases can the source of individual flows be found. Most of the Martian plains are also thought to be composed of lava flows. Extensive flows in these plains have been identified on high-resolution Viking pictures. Their dimensions, widespread distribution, and morphology indicate that they are flood basalts.

The morphology and dimensions of Mercury's plains resemble those of the lunar maria and Martian plains, indicating they are composed of a material that was emplaced in a highly fluid condition. No large volcanic mountains are evident, as are seen on Earth and Mars, which further suggests they were erupted in large volumes from widely distributed fissures. From this evidence scientists have inferred that the plains are the Mercurian equivalent of flood basalts.

Thermal history studies indicate that early in Mercury's evolution conditions were ideally suited for the generation of flood basalts on a global scale. The formation of Mercury's large iron core, together with radioactive heating and heat produced by despinning, would have led to extensive melting, including most of the outer silicate (rocky) layer. The amount of melting produced by these processes should have been sufficient to expand the planet by about 30 kilometers, extensively fracturing the thin crust. Expansion generates tensional stresses on a global scale in a planet's outer solid shell; these stresses produce fractures called normal faults, which allow lavas to migrate to the surface. Thus, Mercury's largely molten interior would have provided a ready source for the lavas, and the fractures would have provided the avenues along which the lavas could travel to the surface. These events would have occurred early in the planet's history, during the period of heavy bombardment, and they may have led to the formation of the intercrater plains.

The smooth plains may have been simply a continuation of these volcanic flooding events, extending into a period when the impact rate had declined significantly and the interior was beginning to solidify. Tectonic studies show that the smooth plains formed while the crust was under compression. Compressive stresses would have tended to close off the source region for lavas. But these younger plains cover a much smaller area than intercrater plains, and they are restricted to the vicinity of large impact basins. Large basin-forming impacts produce extensive fractures beneath and surrounding the basins; the smooth plains may have been extruded along these fractures while the lava's fluid pressure still exceeded the compressive pressure in the crust. As the planet continued to cool and contract, the compressive pressure would eventually have exceeded the fluid pressure of the lava, and these fractures would have sealed to completely shut off the volcanism.

Chapter 8

The Shrunken Planet

Crustal movements have deformed, to one degree or another, the surfaces of almost all solid bodies in the Solar System. Mercury is no exception. In fact, Mercury has experienced a deformational history unlike that of any other planet or satellite. This deformation appears to have been caused mainly by heating and cooling.

Crustal deformation, called tectonism, results from stresses produced in the outer layers of a planet or satellite. These stresses are products of either thermal or mechanical processes, or some combination of both. They form characteristic structures (faults or folds) that reveal the nature and direction of the responsible stresses.

To discover the mechanism that caused a particular deformation, three problems relating to tectonism must be solved. First, the type of deformation must be identified. This is accomplished by studying its morphology and the way in which it displaces or deforms the surface. Second, the distribution of the structures across the surface and the time at which they formed must be determined. This procedure involves plotting the structures on a map and observing their age relative to other landforms. (For instance, if a fault cuts across a crater or a certain terrain, then it obviously is younger than these features. Conversely, if a fault is disrupted by a crater or is partially covered by a certain terrain, then it is older than these features.) Finally, the nature of the stresses (compression versus tension) and their direction must be inferred from the types of structures and their pattern. This latter charac-

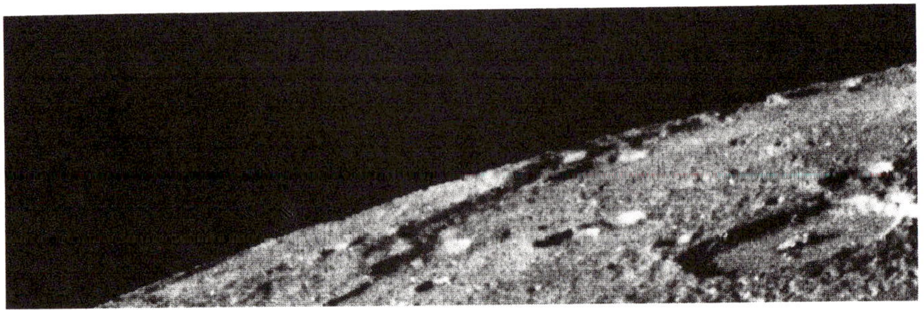

The cliff near the limb of Mercury is a large lobate scarp caused by compressive stresses. (Courtesy NASA.)

teristic is often called the tectonic framework. Not surprisingly, Mercury has a tectonic framework that is unique in the Solar System.

Lobate Scarps

One of Mariner 10's most important discoveries was its observation of long, sinuous cliffs or scarps that cut Mercury's surface for hundreds of kilometers. These cliffs have been termed lobate scarps because their faces feature a characteristic rounded and lobed appearance. Their lengths vary from about 20 kilometers to more than 500 kilometers, and their heights from a few hundred meters up to about 3 kilometers. Earth-based radar measurements have determined heights of about 700 meters and widths of about 70 kilometers for some of these scarps.

Individual scarps often cut through or transect several different terrains, including craters, intercrater plains, and smooth plains. Landforms are displaced where scarps transect them. These relationships indicate that lobate scarps are faults. But precisely what type of fault are they?

There are three basic types of faults: normal, thrust, and transverse. Each is formed by a different stress regime. A convenient way to visualize these regimes is to consider

The lobate scarp traversing Mercury's surface for about 400 kilometers from A to B is called the Discovery scarp. It is one of the largest scarps viewed by Mariner 10, reaching more than 2 kilometers high. (Courtesy NASA.)

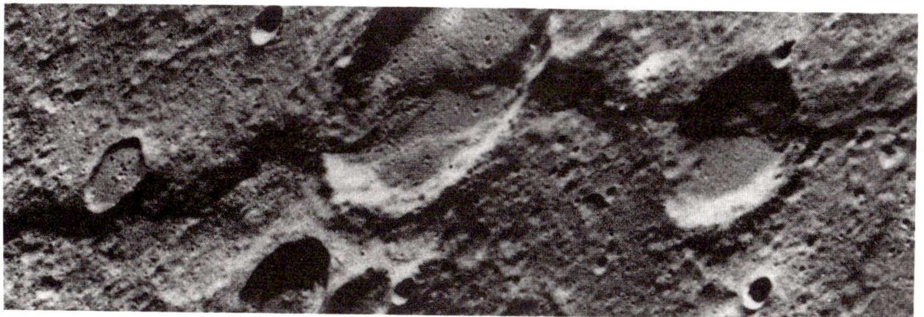

This high-resolution photograph of the Discovery scarp was taken by Mariner 10 on its third encounter with Mercury. (Courtesy NASA.)

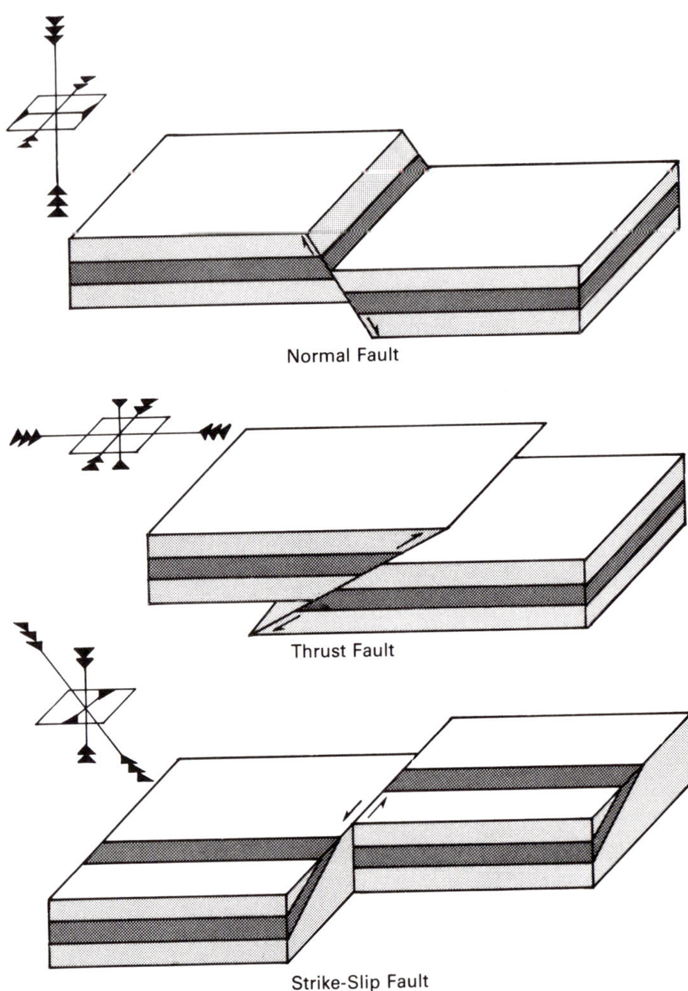

The three basic types of faults. Normal faults caused by tension are rare on Mercury, and strike-slip faults have not been observed. Thrust faults, which are quite common, formed Mercury's lobate scarps.

the stresses as directed along three principal axes at right angles to each other. One axis is the maximum principal stress, another is the intermediate principal stress, and the third is the minimum principal stress. The orientation of these principal stress axes with respect to the surface determines the type of fault that forms. Faults form at an angle between the maximum and minimum stress axes and parallel to the intermediate stress axis.

When the crust of a planet is stretched or pulled apart, it is in tension. The lateral or confining pressure is eased, and the weight of the rocks acting vertically exerts the greatest stress on the crust. Therefore, the maximum stress axis lies perpendicular to the surface, and the minimum and intermediate stress axes are parallel. If the maximum stress exceeds the strength of the rocks, then the crust will fracture. It will break along a fault plane inclined at a steep angle between the maximum and minimum stress axes and parallel to the intermediate stress axis. One block will move downward along a fault plane sloping toward the down-dropped block. This type of fault is known as a normal fault. It is also called a tension fault because it forms when the crust is under tension. Another synonym is "gravity fault," because the maximum stress is vertical and due to the force of gravity on the rocks. This type of faulting often forms troughs where a section of the crust slides downward between two oppositely facing normal faults. This type of fault trough is termed a graben. The fractures on the floor of the Caloris basin are grabens. Grabens are also common on the Moon and Mars.

If the crust of a planet pushes together, then it is in compression and the stress field reverses. Now the maximum stress axis parallels the surface and the minimum stress axis is vertical. Again, the crust breaks along a fault plane inclined at an angle between the maximum and minimum stress axes and parallel to the intermediate stress axis. In this case, however, one block is pushed or thrust over another block along a gently sloping fault plane that dips beneath the overthrust block. This type of fault is known as a thrust fault.

The third type of fault forms when both the maximum and minimum stress axes parallel the surface and the intermediate stress axis is perpendicular. Again, the fault plane forms at an angle between the maximum and minimum stress axes, but in this case the fault plane is vertical and one block slides past the other with little or no vertical displacement. This type of fault is called a transverse fault.

Thus, normal faults result from tension that pulls the crust apart and causes crustal lengthening. Thrust faults are due to compression that pushes the crust together, causing crustal shortening. In transverse faulting two blocks slide laterally by each other and the surface area is conserved; that is, neither crustal lengthening nor crustal shortening occurs.

The surface expressions of these three types of faults differ from each other. Normal faults or grabens form linear or slightly curved scarps or troughs that vertically displace the topography they cut; their scarps are generally steep with sharp crests. Transverse faults are quite straight, show little vertical relief, and laterally displace landforms they transect. Thrust faults vertically displace the surface, but their crests are rounded because the fault plane dips beneath the overthrust block, producing an overhang that is unstable and sags under its own weight. The surface trace of a thrust fault is usually sinuous, be-

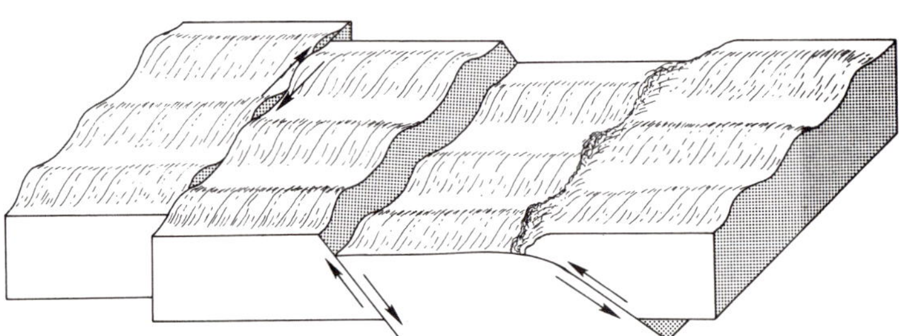

The three types of faults exhibit different surface expressions. Strike-slip faults (left) show straight fault traces that horizontally offset topographic features. Normal faults (center) are relatively straight or gently curved, displaying vertical displacements of topography. Thrust faults (right) feature a sinuous fault trace, with vertical displacements of topographic features. Mercurian lobate scarps show this type of surface expression.

cause the fault plane slopes at a low angle and forms a meandering path as it cuts across terrain of different elevations. Thrust faults also push one portion of the surface over another so that landforms are shortened.

The surface expression of Mercury's lobate scarps is exactly that expected of thrust faults. They are sinuous scarps with rounded crests that in some cases greatly distort the craters they transect. One outstanding example of a distorted crater is Guido d'Arezzo, whose rim is cut by a

The fault that displaces the crater rim in the center of this photograph is the Vostok scarp. (Courtesy NASA.)

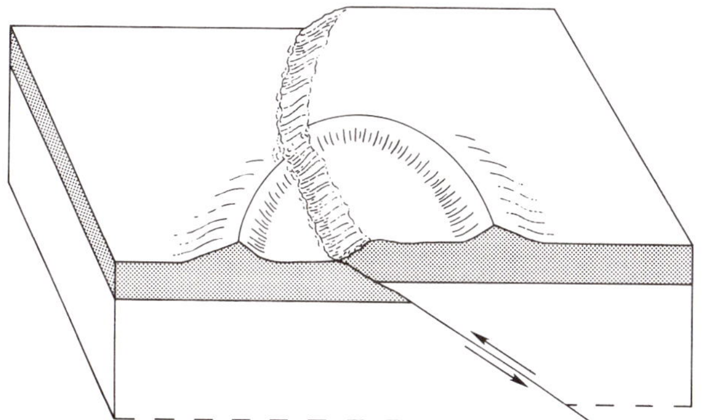

This diagram illustrates how the Vostok thrust fault has displaced the rim of the Guido d'Arezzo crater so that it appears to be offset horizontally. The crater's diameter has been shortened by thrusting of its right part over the left part. As a result, the rims do not join up and are offset.

lobate scarp named Vostok. The northeastern part of this crater has been thrust over the southwestern part, causing a shortening of the crater's diameter and a 10-kilometer horizontal offset of its rim. Therefore, both the morphology of the scarps and the surface displacements indicate that Mercury's lobate scarps are thrust faults formed by compression.

Mercury's surface is remarkably free of structures indicative of tensional stresses. Only the grabens on the floor of the Caloris basin, and the valleys (probably grabens) and jumbled surface of the hilly and lineated terrain, are due to tensional stresses. But both these structural provinces reflect the direct or indirect effects of the Caloris impact. The grabens on the floor of Caloris resulted from a local uplift following the impact, while the hilly and lineated terrain was caused directly by seismic waves generated by the impact and focused at the antipodal region. The present structure of Mercury is therefore dominated by thrust faults caused by compressive stresses.

Distribution and Age of Lobate Scarps

Lobate scarps are found in all regions and on all types of terrain viewed by Mariner 10. They are more or less evenly distributed over the equatorial, mid-latitude, and polar regions. Their orientations reflect no obvious preferred direction, but they seem to trend somewhat more frequently in directions between N 45°W and N 45°E. The broad occurrence of lobate scarps on the half of Mercury imaged by Mariner 10 suggests that they are scattered across the planet's entire surface. If so, then the entire planet suffered a period of global compression. This hypothesis implies a net decrease in surface area since the onset of compression.

The amount of decrease in surface area can be estimated from the total length and height of the scarps, and by assuming an inclination angle (the slope) of the fault planes. A simple trigonometric relationship yields the amount of horizontal displacement or crustal shortening that would result if the fault plane inclination and amount of vertical displacement of a thrust fault are known. The decrease in surface area along a thrust fault is simply the length of the fault times the amount of vertical displacement divided by the tangent (a trigonometric function) of the fault plane inclination. The lengths of the faults are easy to measure, and the amount of vertical displacement is probably similar to the heights of the scarps. The average height of the scarps is about 1 kilometer. The inclinations of the fault planes are more difficult to estimate, but they probably range between about 25 and 45 degrees. These values seem to be consistent with the inclinations estimated from the amount of crater rim displacements on Mercury.

Using these values and assuming that the imaged portion of Mercury is representative of the planet as a whole, then the total decrease in surface area has been between about 63,000 to 130,000 square kilometers. This amount of crustal shortening could be explained if Mercury's diameter has decreased by about 2 to 4 kilometers since the onset of compression. In other words, Mercury has apparently shrunk. No evidence suggests that any other solid planet or satellite has experienced this amount of shrinking.

When did this decrease in diameter occur? The age of lobate scarps relative to other surface features can be determined by transection relationships. In the uplands, the scarps cut across the intercrater plains and all relatively old degraded craters. Nowhere do the intercrater plains partially cover or embay the lobate scarps. Craters that disrupt scarps are all relatively young, fresh-appearing ones. These relationships indicate that the onset of scarp formation (compression) occurred after the intercrater plains were deposited and at a time when the period of heavy bombardment was declining. Lobate scarps are somewhat fewer on the younger smooth plains than in the older uplands. This suggests that some of the earlier formed scarps were buried by the smooth plains. Therefore, most smooth plains were probably erupted slightly after the onset of compression. These observations indicate that Mercury's diameter began to decrease after the formation of intercrater plains and slightly before the majority of smooth plains were deposited. Since lobate scarps cut the smooth plains, the decrease must have continued after the formation of smooth plains. Even today the crust may be under compression.

Other Structures

In addition to the lobate scarps, another less prominent set of linear features may have a tectonic origin. They consist of linear portions of crater rims and linear, ridge-like structures. These lineations trend in two main directions, northeast and northwest, with a weaker north-south direction. They may represent an ancient fracture pattern formed early in Mercury's history. The valleys in the hilly and lineated terrain trend in the two main directions. They may have formed along fractures of this system when the Caloris seismic waves disrupted the area. If so, the fracture system was present before the Caloris impact.

The linear portions of crater rims may have formed by slumping or preferential excavation along zones of weakness associated with such a preexisting fracture pattern. These linear rims occur on the more degraded, and therefore older, craters. This pattern indicates that the fracture system was present before the older craters and therefore was established quite early in Mercurian history.

Origin of the Tectonic Framework

Two mechanisms have probably acted to deform Mercury's crust. One involves mechanically produced stresses, and the other thermally induced stresses. Mercury's rotation rate was discussed in chapter 3. Recall that the planet may have rotated much faster just after its forma-

The arcuate scarp near the center of this photograph occurs in Mercury's south polar region. (Courtesy NASA.)

tion and was subsequently slowed by tidal forces until it was captured into the present 3:2 spin-orbit resonance. During this epoch of rapid rotation, centrifugal forces would have produced an asymmetrical shape with flattened poles and a bulging equator. As Mercury slowed and the centrifugal forces decreased, its shape would have become more spherical.

In changing from a flattened sphere (oblate spheroid) to a sphere, the bulging equatorial regions would have contracted and the flattened polar regions would have expanded proportionately. This process would have produced a stress pattern leading to a unique tectonic framework: the contracting equatorial regions would have produced east-west directed compressional stresses, with the expanding polar regions inducing north-south directed tensional stresses. In the equatorial regions, therefore, north-south-trending thrust faults would have formed, and in the polar regions, east-west trending normal faults.

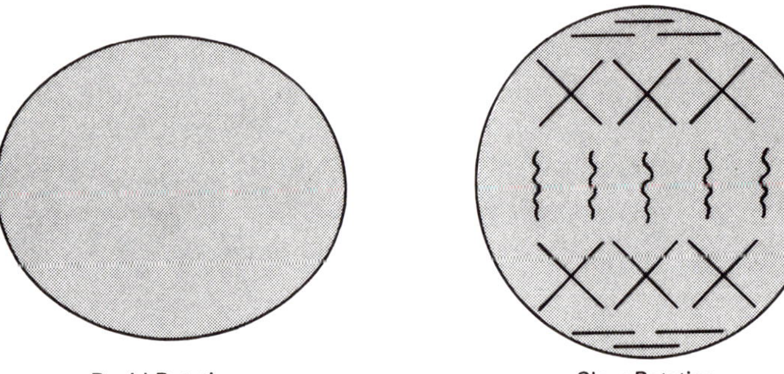

If Mercury initially rotated much more rapidly than later, it would have possessed flattened poles, as shown at left. As Mercury's rotation slowed, it would have become more spherical, causing stresses above the fracture limit. This process would have resulted in the pattern shown at right, with east-west normal faults in the polar region, northeast-southwest and northwest-southwest strike-slip faults in the mid-latitudes, and north-south thrust faults in the equatorial region.

In the mid-latitudes, this interplay of stresses between the equatorial and polar regions probably led to a system of northwest and northeast trending transverse fractures.

This mechanism, however, cannot be the main cause of Mercury's present tectonic framework, because thrust faults are just as abundant in the polar regions as they are in the equatorial regions, thrust faults have a more or less random direction, and no normal faults have been recognized in the polar regions. However, the northwest and northeast set of ancient lineations may be the remains of a transverse fracture system caused by planetary despinning. The ancient age of these lineations suggests that if despinning occurred, it did so quite early in Mercury's history, probably before the formation of intercrater plains and many of the older craters.

Most substances will expand when they are heated and contract as they cool. Rocks and iron are two such materials. Thermal history studies indicate that early in its history Mercury experienced extensive heating and melting, followed by cooling (see chapter 4). The early heating

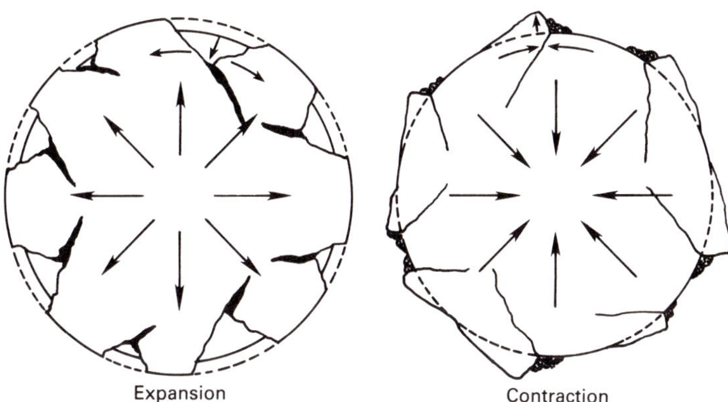

Planetary expansion (left) will cause a surface to stretch, producing tension and normal faults. This process may have occurred on Mercury early in its history. During planetary contraction (right), the crust would be compressed, forming thrust faults (lobate scarps). This type of fault is common over most of Mercury's surface.

phase would cause Mercury to expand, and as it cooled the planet would contract. The widespread (probably global) distribution of thrust faults strongly suggests that the entire planet contracted. This contraction appears to have begun after the intercrater plains were deposited and slightly before smooth plains deposition. Contraction, therefore, began rather late in the active part of the planet's history.

Initially, it was thought that the solidification of Mercury's large iron core was responsible for the contraction and consequent formation of thrust faults. But thermal history studies show that only the cooling of the outer rocky mantle and crust would have been needed to account for the amount of decrease in radius estimated from the thrust faults. In fact, a complete or even extensive solidification of Mercury's large core would have caused a much greater radius decrease than can possibly be accounted for by the thrust faults. These factors imply that the core is still largely, but not completely, molten, which is consistent with the interpretation of Mercury's magnetic field. Thus, the planet's system of lobate scarps is probably the result primarily of the outer rocky layers cooling after an intense period of melting and expansion.

Early Global Expansion

All the terrestrial planets, as well as most outer planet satellites, were heated during and just after their formation. In most planets this heat derives mainly from the radioactive isotopes of uranium, thorium, and potassium. As these isotopes decay, they produce heat, and the amount of heat depends on their abundance.

Cosmochemical studies of solar nebula condensation and planetary accretion have predicted that Mercury should be depleted in potassium, but that it should retain a considerable abundance of uranium and thorium. A uranium abundance of only forty-four parts per billion is sufficient to raise the temperature above the melting point of iron and some silicate minerals. At least this amount of uranium was probably present in Mercury. The cosmochemical models also predict that the ratio of thorium to uranium was about four to one. These abundances of uranium and thorium probably led to the melting of Mercury's iron content and some silicate minerals. Because iron is much denser than silicates, it would migrate toward the center of the planet to form a core.

The movement of iron toward the center of Mercury also would have produced heat. The amount of heat produced in this way depends on the abundance of iron. Mercury contains about 70 percent by weight of iron—more than any other planet or satellite in the Solar System. The movement of this much iron toward its center would have raised its temperature by about 700° C.

But these processes were not the only sources of heat. As Mercury's rotation slowed from a more rapid rate, a portion of the tidal friction that caused the slowing would be converted to heat. The amount of heat generated in this way may have increased the planet's internal temperature by another 100° C. The accretion process itself could also have added heat to Mercury. We have already seen that a portion of the kinetic energy of impact takes the form of heat. The extremely high rate of impact during Mercury's accretional stage could also have heated the planet's outer portion.

This combined heating by accretion, tidal despinning, radioactive decay, and core formation was sufficient to almost completely melt Mercury, to an extent at least equal to, and probably greater than, that of the other terrestrial planets. This amount of heating caused the planet to expand significantly; it has been estimated that its diameter increased by about 30 to 35 kilometers. Heat loss by conduction and convection in the outer layers would have produced a thin, solid crust that was subjected to considerable tensional stresses. The amount of crustal lengthening resulting from this expansion would have been about 1 million square kilometers. The crust would fracture extensively to form normal faults. These faults could have tapped the relatively shallow molten interior and given rise to extensive volcanism on a global scale.

The only surface evidence for such a period of global expansion is the intercrater plains. These plains are probably volcanic flood basalts extruded along fractures formed during global expansion. The lack of observed normal faulting in Mercury's uplands may be the result of volcanism and impact cratering. Crustal expansion and tensional faulting may have been nearly complete by the time the oldest observed intercrater plains were erupted. These eruption products may have covered the source fractures, just as the lunar mare lavas have covered theirs. The expansion must have occurred during the period of heavy bombardment. Impact cratering (both primary and secondary) could also have obliterated many of the fractures.

Our current understanding suggests that thermal processes were largely the cause of Mercury's crustal deformation. A period of early heating and global expansion probably produced tension fractures, which gave rise to intercrater plains volcanism. This period was followed by cooling global contraction that formed the lobate scarps or thrust faults. All these events happened early in Mercury's history.

☿

Chapter 9

Through the Ages

The history of a solid planet or satellite derives from measurements of its physical properties and environment, studies of its surface features, composition, and properties, and theoretical modeling of various processes. For Mercury, this information remains quite limited. We have seen less than half the planet and only at moderate resolution. Almost nothing is known about its surface composition. We have had only glimpses of its magnetic field, and theoretical modeling is still incomplete. As a consequence, there is considerable uncertainty concerning Mercury's history. Nevertheless, from these limited data our current understanding of the planet is sufficient to attempt a reconstruction of its history.

Surface Ages

As in any history, ages of events must be known to place them in their proper order. Two types of age measurements can be made for planetary surfaces: absolute and relative. Absolute ages yield the actual times at which surfaces formed. They can only be measured accurately by complex laboratory analyses of returned rock samples. Relative ages date surfaces or events with respect to each other. They give no information about the actual times these events took place, but they do place them in a correct relative sequence. Relative ages are determined by well-established geologic methods involving impact crater abundances and superposition and transection relationships.

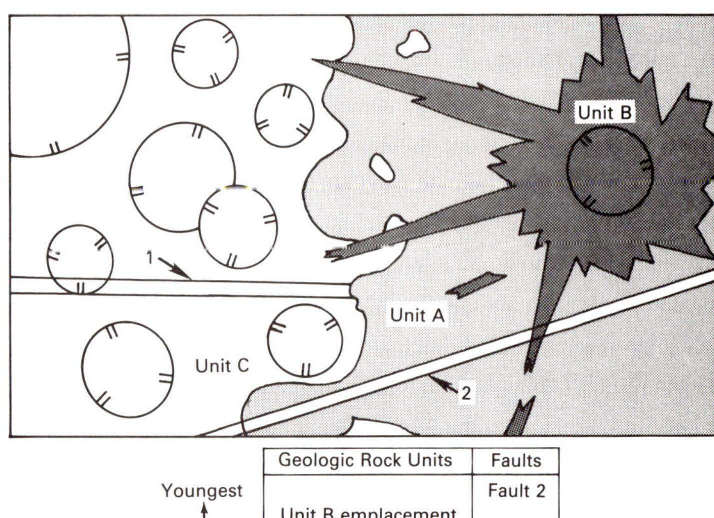

	Geologic Rock Units	Faults
Youngest		Fault 2
↑	Unit B emplacement	
	Unit A emplacement	
		Fault 1
↓	Unit C emplacement	
Oldest		

The ages of various surface units can be dated relative to each other by superposition or cross-cutting relations. In this example, the crater ejecta of Unit B was deposited on Units A and C and therefore is younger than these units. Unit A overlaps and is less cratered than Unit C, and so is younger than Unit C. Fault 1 cuts Unit C but is overlapped by Unit A. It is therefore younger than Unit C but older than Unit A. Fault 2 cuts all units and is therefore the youngest feature in the area. The table lists the features in order of increasing age.

The relative ages of various Mercurian terrains and tectonic structures viewed by Mariner 10 are fairly well known. Terrains have been dated relative to other terrains or landforms by the number of impact craters on their surfaces and whether or not they partially cover or embay other terrains or landforms. If one surface has accumulated more primary impact craters than another surface, then it must be older. A terrain that partially covers another is younger than the one it covers. Faults can be dated relative to other features by their transection relationships.

For instance, if a fault cuts through and displaces a surface or landform, then it must be younger than these features. On the other hand, if a fault is partially buried by a terrain or disrupted by a crater, then it predates these landforms.

Craters can be dated relative to each other by superposition relationships. Obviously, if a crater or its ejecta deposit overlies other craters, it is younger than those craters. But craters far removed from each other also can be dated from their states of preservation. A highly degraded crater is older than a similar-sized crater with sharp, fresh-appearing features; it is important to compare craters of similar sizes when gauging degradational states, because small craters degrade faster than large craters.

These relative age-dating techniques have been used to establish a sequence of geologic events for Mercury. The intercrater plains span a variety of ages, all of which are contemporaneous with the period of heavy bombardment. Thus, the heavily cratered uplands and intercrater plains are the oldest observed surfaces on the planet. The system of lineations also probably formed during this time. Lobate scarps or thrust faults began to form after the intercrater plains had been deposited. The next major event was the Caloris basin impact and the formation of hilly and lineated terrain antipodal to it. This was followed by the emplacement of the majority of smooth plains in and around the Caloris and other large impact basins. Lobate scarps continued to form after the majority of smooth plains were emplaced, but when this event ended is not known.

The formation of smooth plains was the last major event to occur on Mercury. At this time, the frequency of impacts had declined significantly, and since then the surface has slowly been accumulating impact craters up to the present. But when in geologic history did these events take place? Were they spread out over a long period, or were they compressed into a short interval? Only absolute age dating can reveal the answers.

high Impact Rate low	
Formation of Crater Degradational Classes	5 - 3 2 1
Plains Formation	Intercrater Smooth
Caloris Related	Caloris Impact, Hilly and Lineated Terrain
Scarp Formation	
Thermal History	Melting & Expansion Cooling & Contraction

Time ⟶

This schematic diagram summarizes the sequence of events in Mercury's history. The events are listed in the left-hand column, and the duration or magnitude of the event is indicated in the right-hand column. Time increases from left to right as indicated by the arrow at bottom. Most of the events probably occurred during the first billion years of the planet's existence. Early in Mercury's history, the impact rate was extremely high but declined rapidly. During this time, craters that are now degraded (5–3) were being formed. Also at this time, the planet was experiencing melting, expansion, and formation of intercrater plains. Near the end of heavy bombardment, when the impact rate was lower, craters (2) formed that are now less degraded. At this time, the Caloris basin and the hilly and lineated terrain formed, and the planet began to cool and contract. The smooth plains were then formed, and planetary contraction continued to create the lobate scarps. Since then, only rare impacts have occurred, forming the freshest craters (1).

Ideally, scientists would like to date events on an absolute time scale, but this is not always possible. There are two ways of obtaining absolute ages of planetary surfaces. One involves dating a returned rock sample by radioactive techniques, and the other measures impact crater abundances (if the impact rate is known). Radioactive dating is by far the most accurate and desirable method of dating, but as yet rock samples are available only from the Earth, the Moon, and meteorites (one type of meteorite may have come from the surface of Mars).

Most rocks have solidified from a molten state (sediments are an exception). They may have been melted in the interior of a planet, or they may have been melted at the surface by impacts. The time a rock solidified from a melt is determined from the radioactive decay of certain unstable isotopes such as uranium 238 and rubidium 87. When such isotopes decay, they produce more stable daughter isotopes. The unstable isotopes decrease in abundance with time, and more stable daughter isotopes increase in corresponding number. The age of the rock can be calculated from the known decay rate of the isotope and from the relative proportions of parent and daughter isotopes. This method has been used to date lunar and terrestrial rocks and meteorites. Of course, we do not yet have samples from Mercury on which to perform such measurements.

The situation is not completely hopeless, however. If the impact rate (the number of impacts per unit time) is known, then the time a surface formed can be estimated from the abundance of superposed impact craters. Only for the Moon, however, do we have a good estimate of the past impact rate and how it varied with time. The ages of different lunar surfaces have been determined from samples returned by the Apollo astronauts. These surfaces span a range of ages from about 4 to 3 billion years. By counting the number of craters that accumulated on surfaces of different ages, we can determine the number of craters that formed during different intervals. From this relationship we derive the impact flux, or the number of impacts per unit area as a function of time. By determining crater abundances at other localities, it has been possible to estimate absolute ages of lunar surfaces not sampled by the Apollo astronauts. The youngest lunar lava flows dated in this way are about 1 billion years old.

The lunar impact flux curve shows that the impact rate was extremely high from about the time of lunar formation (4.6 billion years ago), declining rapidly until about 3.8 billion years ago. From 3.8 billion years ago to the present, the impact rate has been low, with only a gradual decline. The high but rapidly declining impact rate between 4 and 3.8 billion years ago marked the period of heavy bombardment represented by the heavily cratered highlands. As discussed in chapter 6, this period of heavy bombardment was caused by a family of impacting objects that, at least on the Moon, became extinct 3.8 billion years ago. The low, gradually declining impact rate between 3.8 billion years ago and the present represents another family of impacting objects that apparently is still colliding with the terrestrial planets.

Whether or not the lunar impact flux was the same for other terrestrial planets is uncertain. It depends on one's assumptions about the origin of the impacting objects. As previously discussed, the crater size/frequency distributions for the heavily cratered terrains on the Moon, Mercury, and Mars are the same. This correspondence indicates that the same family of impacting objects was responsible for the period of heavy bombardment on all the terrestrial planets. Because this period ended 3.8 billion years ago on the Moon, scientists are fairly confident that, no matter what the origin of the objects, the heavily cratered surfaces are extremely ancient and date from the earliest period in Solar System history. It is unlikely that comets or asteroids were significant contributions to the period of heavy bombardment (see chapter 6). If the impacting objects were largely remnants left over from the accretion of the terrestrial planets, then they would have been swept up by these planets and become extinct at roughly the same time on the Moon and Mercury. If this were the case, then the heavily cratered terrains on Mercury are probably older than about 4 billion years.

The visible lunar maria are younger than about 3.8 billion years and are characterized by a low crater abundance that has a different size/frequency distribution from the heavily cratered terrains. This indicates that the maria were formed after the end of heavy bombardment, when this family of objects had become extinct. On Mercury, however, the smooth plains surrounding the Caloris basin are marked by a crater abundance about five times greater than the lunar maria but still much less than the uplands. Furthermore, the size/frequency distribution of these craters equals that of the heavily cratered terrains on the terrestrial planets. These factors suggest that the smooth plains on Mercury formed near the tail end of heavy bombardment, but at a time when this family of objects still dominated the impact flux. If the period of heavy bombardment ended on Mercury at the same time it did on the Moon, then the Mercurian smooth plains may be significantly older than the lunar maria, perhaps about 3.8 or 3.9 billion years old. If so, then all of Mercury's major geologic events occurred during the first 700 or 800 million years of Solar System history. On the Moon, the major events spanned at least the first 1.5 billion years.

The History of Mercury

Some uncertainty remains concerning the origin of Mercury's plains, but the current evidence favors a volcanic origin for both the smooth and intercrater plains. The lobate scarps are almost certainly thrust faults caused by crustal compression. Focused seismic waves from the Caloris impact probably formed the hilly and lineated terrain. The relative ages of these major features are fairly well established, but the actual times they formed are unknown. We can be sure, however, that Mercury's surface is extremely ancient and dates from the earliest period of Solar System history. Thermal history models also seem to be consistent with these interpretations of the surface features and their relative ages. The following capsule history of Mercury is based on these interpretations.

Four and a half billion years ago, Mercury formed by the accretion of dust and gas in the hot inner regions of the primitive solar nebula. Its formation in this part of the nebula resulted in a high concentration of iron throughout the planet. Just after formation it was probably rotating more rapidly than at present. Also at this time, Mercury was subjected to an intense bombardment by objects that may have been remnants from accretion.

Heat from the decay of radioactive elements, possibly combined with the heat of accretion, began to melt the planet. Iron began separating from silicates and started migrating toward the center to eventually form a large core. Heat caused by this migration raised the temperature an additional 700° C. Mercury's rotation was slowing down, which also increased the internal temperature and fractured its thin crust.

As the temperature rose, Mercury became increasingly molten and began to expand. At this time, the magnetic field probably developed. As melting and expansion proceeded, the thin, solid crust was torn by fractures caused by the expansion. These fractures tapped the shallow molten interior as meteorites continued to rain onto the surface. On all parts of the planet enormous eruptions of lava issued from the fractures and poured onto the surface, forming the intercrater plains. The lavas inundated older impact craters and, in turn, were cratered by the intense bombardment. These lavas were relatively depleted in iron and titanium because these elements had earlier migrated to deeper levels during core formation. This period of volcanism may have lasted a few hundred million years.

As Mercury began to cool and its crust thickened, the eruptions became somewhat less frequent. As cooling and crustal thickening proceeded, the planet began to contract, placing the crust in compression and initiating thrust faults that formed lobate scarps. Planetary despinning may have been largely completed, but it could have aided in the formation of some thrust faults. By this time, the impact rate had declined from a previously higher level.

Just after the onset of planetary contraction, a large impact created the Caloris basin and generated seismic waves focused at the antipodal region. These seismic waves severely disrupted the crust to form the hilly and lineated terrain. The Caloris impact may have occurred about 4 billion years ago.

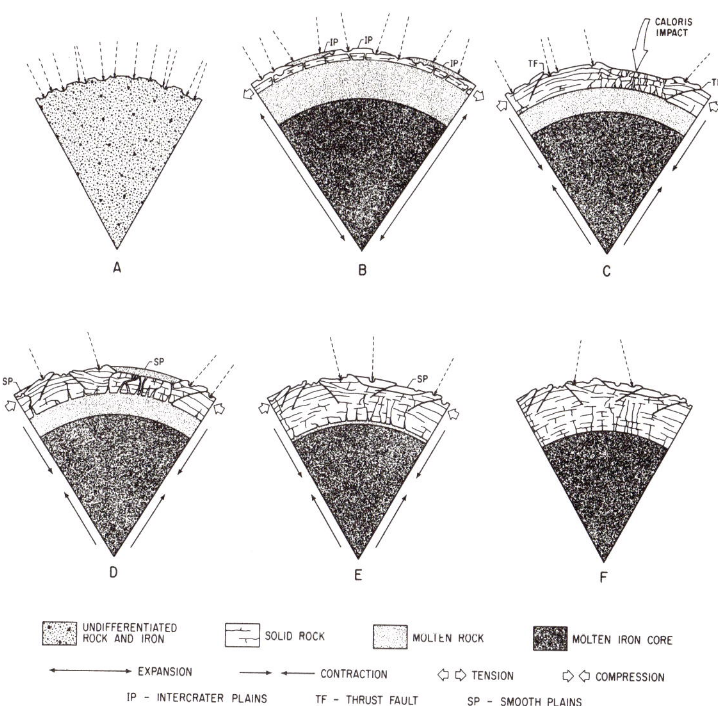

A further representation of Mercury's possible history. In the first diagram, Mercury has just formed, with a large percentage of iron uniformly distributed throughout the planet. In Diagram B, the large iron core has formed, causing large-scale melting of the rocky mantle, global expansion, and tensional fracturing of a thin, solid crust. Volcanic eruptions along these fractures produce the intercrater plains. In Diagram C, the cooling mantle begins to solidify, causing global contraction, compressive stresses, and thrust faults. About this same time, the Caloris impact occurs, causing the hilly and lineated terrain on the planet's opposite side. In Diagram D, new eruptions fill the Caloris basin and other areas with smooth plains near the end of heavy bombardment. In Diagram E, further cooling and contraction produce compressive stresses strong enough to close off the volcanic conduits. Finally, in Diagram F, cooling and contraction are nearly complete, and the planet becomes inactive except for occasional impacts that scar the surface. Mercury may have a solid inner core that is not shown in these illustrations.

Soon after the Caloris impact, lavas erupted onto the Caloris basin floor and surrounding region, and onto the floors of older impact basins. These lavas—creating the smooth plains—were probably no more than a continuation of intercrater plains volcanism, which had begun to ebb as Mercury cooled and contracted. They were concentrated in and around large impact basins because it was here that the relatively thin crust was intensely fractured by the impacts and could be more easily breached after the planet began to contract. The Caloris lavas are darker than intercrater plains lavas, probably because they originated from a deeper layer in the planet's mantle where iron and titanium were more concentrated. At this time, the floor of the Caloris basin subsided, causing its surface to become wrinkled. Soon afterward it rose, producing the radial and concentric fracture pattern.

By the time the smooth plains lavas were erupted, the intense bombardment had decreased significantly and was beginning to come to an end. Mercury continued to cool, causing further contraction, additional thrust faults, and thickening of the lithosphere (the outermost rigid part of a planet). Finally the lithosphere became so thick and the compression so strong that molten material could no longer reach the surface, and volcanism ceased. The last major eruptions may have ended as early as 3.8 or 3.9 billion years ago.

After these last major eruptions, the period of heavy bombardment ended. Mercury continued to cool and contract, causing more thrust faults, but when contraction ceased is not known. Since the end of heavy bombardment, the planet's surface has been impacted less frequently by comets and asteroids.

Comparison with Other Terrestrial Planets

The history of a planet is largely governed by its internal temperature. The hotter its interior, the more active are both its interior and surface. If a planet is able to retain its internal heat, then its activity will be prolonged. It takes longer for a large planet to lose its heat than for a small planet. Therefore, large planets generally remain active longer than small planets. Other effects, such as tidal heating, can also prolong and maintain high internal temperatures, as occurs in Jupiter's satellite Io.

The Earth is the largest and most active terrestrial planet. It has retained relatively high internal temperatures thoughout its history, and consequently its surface and interior have remained active and dynamic up to the present. Earth's outer rigid layer—the lithosphere—is broken up into large plates that move relative to each other over a partially melted zone called the asthenosphere. Active volcanism occurs almost continuously along the boundaries between plates and over "hot spots" called mantle plumes. All the ocean floors (about 60 percent of Earth's surface) have been created during the last 5 percent of geologic history by a plate tectonic process called seafloor spreading. Thus, the Earth is extremely active and is constantly changing on a geologically short time scale.

Mars is about half the size of Earth and consequently has retained less of its heat for a shorter period. Its heavily cratered southern hemisphere still retains the signature of the period of heavy bombardment. Plate tectonics has not been active on Mars, and therefore most of its surface is quite ancient. Volcanism, however, has been extensive, and at least in one area (Tharsis), it has persisted throughout most of geologic history. The most recent eruptions may have occurred less than 100 million years ago. Martian internal activity has been much less extensive and long-lived than on Earth.

The Moon is the smallest terrestrial-like body. Consequently, it lost much of its heat quite early. About 80 percent of its surface is ancient, and its geologic activity has been short-lived. The last major event in lunar history

This digital airbrush mosaic of Mercury provides a mapping base for future mission planning, and is itself a new experiment in planetary cartography. By computer encoding of the Mercurian maps, the planet's surface can be portrayed in different map projections. (Courtesy of U.S. Geological Survey, Flagstaff.)

was volcanic flooding to produce the mare-filled impact basins. Most of this internal activity ended about 2.5 to 3 billion years ago.

In many respects, Mercury's history is quite different from those of the other terrestrial planets. Most of these differences can be explained by the direct or indirect effects of the formation of Mercury's enormous iron core. The eruption of the intercrater plains lavas was probably the result of extensive melting and crustal expansion, caused by heat that was mostly generated from core formation.

Mercury appears to have been considerably more melted than the Moon, and possibly than the other terrestrial planets as well. This melting may have led to a more thorough differentiation of Mercury's rocks, in which iron- and titanium-bearing minerals are depleted compared to lunar rocks. Under Mercury's more molten conditions and greater gravity, these minerals would rapidly sink to lower levels before the outer liquid layer could solidify. This may account for the higher albedo of Mercury's lavas, because iron- and titanium-bearing minerals darken rocks.

Global contraction and thrust faulting were indirect consequences of core formation, because a significant decrease in Mercury's diameter resulted from the cooling that followed the extensive melting and expansion caused by core formation. No other planet or satellite has experienced this much contraction on a global scale. It has led to Mercury's unique tectonic framework and may have been responsible for an early end to volcanism. As the planet contracted it would not have taken long before crustal compression closed off the conduits up which lavas migrated, preventing further volcanism. If the period of heavy bombardment ended the same time on Mercury as on the Moon, then Mercury's active history may have ended about 3.8 or 3.9 billion years ago, or over 1 billion years earlier than that of the Moon. Mercury's intense heating and melting may thus have been indirectly responsible for an early end to its active history.

The characteristic that truly sets Mercury apart from any other body in the solar system is its enormous iron core. Mercury's unusual history appears to have been largely governed by this core. Other planets may have had more active and prolonged histories, but only Mercury has been so profoundly affected by its internal constitution.

Epilogue

Our knowledge of Mercury has been advanced a thousandfold by the people who conceived, built, and flew the Mariner 10 spacecraft and by the scientists who analyzed the data returned from the mission. Without this experience, a fundamental gap would have remained in our understanding of the evolution of the terrestrial planets in particular, and of the Solar System in general.

The Mariner 10 mission, however, was never meant to be anything more than a reconnaissance flight—a first brief look that would answer some basic questions about a largely unknown planet. Mariner 10 admirably accomplished this task and more. More than half of Mercury has not been imaged, however, and the other half has only been seen at low resolution. We know little about the planet's surface composition, and its magnetic field is not well understood.

As any good mission should, Mariner 10 posed as many questions about Mercury as it answered. Is the unimaged half of the planet similar to the imaged half, or are there different types of terrains that would significantly alter our current interpretations? How important are the basin ejecta deposits to the formation of the plains? Is the core still largely molten, and if so, what are the heat sources that maintain a molten condition over the age of the Solar System? What is the detailed structure of Mercury's magnetic field, and does it fluctuate in intensity and reverse polarity, as does the Earth's field? What is the surface composition, and how does it differ from that of other terrestrial planets? These are only a few of the questions that must be answered before we can confidently and fully reconstruct Mercury's evolution. Only a Mercury orbiter and eventually landers will be able to provide the answers.

Unfortunately, the prospects for exploration of Mercury in the near future are not bright. Such missions are extremely complex and expensive, and there are no plans to send spacecraft to Mercury during this century. It has been more than thirteen years since Mariner 10 first encountered the planet, and it appears that at least another sixteen to twenty years will pass before our next explorations occur. During this thirty-year period, the Mariner 10 data will remain our only source of detailed information about Mercury. Thus, in many respects, the planet that early astronomers found so difficult and frustrating to observe still remains elusive.

Additional Readings

Briggs, Geoffrey, and Fredric Taylor. *The Cambridge Photographic Atlas of the Planets.* Cambridge: Cambridge University Press, 1982.

Carr, Michael H., ed. *Geology of the Terrestrial Planets.* NASA Special Publication 469. Washington, D.C., 1984.

Chapman, Clark R. *The Inner Planets.* New York: Charles Scribner's Sons, 1977.

Davies, Merton E., Stephen E. Dwornik, Donald E. Gault, and Robert G. Strom. *Atlas of Mercury.* NASA Special Publication 423. Washington, D.C., 1976.

Dunne, James A., and Eric Burgess. *The Voyage of Mariner 10.* NASA Special Publication 424. Washington, D.C., 1978.

Gallant, Roy A. *National Geographic Picture Atlas of Our Universe.* Washington, D.C.: National Geographic Society, 1980.

Greeley, Ronald. *Planetary Landscapes.* London: Allen & Unwin, 1985.

Murray, Bruce C. "Mercury." *Scientific American,* September 1975, pp. 59–68.

Murray, Bruce C., and Eric Burgess. *Flight to Mercury.* New York: Columbia University Press, 1977.

Murray, Bruce C., M. Malin, and R. Greeley. *Earthlike Planets.* San Francisco: W. H. Freeman, 1981.

Ryan, Peter, and Ludek Pesek. *Solar System.* New York: Viking Press, 1978.

Appendix 1

Glossary of Terms

absolute age	The age of a geologic unit measured in years.
acceleration	Change in velocity with time.
accretion	The gradual accumulation of mass, as by a planet forming by the buildup of colliding particles in the solar nebula.
albedo	The ratio of the amount of light reflected from a planet or other body to the amount of light it receives from the Sun.
antipodal point	Point on the opposite side of a planet (i.e., 180 degrees apart on a sphere).
aphelion	Point in its orbit at which a planet is farthest from the Sun.
Apollo program	The American program to land humans on the Moon (1961–72).
asteroid	One of several tens of thousands of small planets ranging in size from a few hundred kilometers to less than 1 kilometer in diameter, with an orbit generally between the orbits of Mars and Jupiter.
asthenosphere	The partially melted or weak layer underlying the outer rigid lithospheric layer of a planet or satellite.

astronomical unit (AU)	The average distance between the Earth and the Sun (about 150 million kilometers).
atom	The smallest particle of an element that retains the properties that characterize that element.
aurora	Light radiated by atoms and ions in the ionosphere, mostly in the magnetic polar regions.
ballistic range	The distance an ejected particle travels before striking the ground.
ballistic trajectory	The looping path that an ejected particle travels while in flight.
basalt	A fine-grained, dark volcanic rock rich in iron and magnesium.
bow shock wave	Boundary where the solar wind encounters the magnetic field of a planet.
breccia	Rock and mineral fragments cemented together.
caldera	A large volcanic depression, often formed by collapse caused by the withdrawal of underlying magma.
cartography	The science and art of expressing graphically, by means of maps and charts, the physical features of a planet's surface.
celestial	Pertaining to the sky.
comet	A small body of icy and dusty matter that revolves about the Sun.
commensurability	The state in which two periods (e.g., rotation and orbital, or orbital periods of two objects) are whole number multiples of each other.
conduction	The transfer of heat from a hotter region to a cooler one by the vibration of atoms.

constellation	A configuration of stars named for a particular object, person, or animal, or the area of the sky assigned to a particular configuration.
continuous ejecta blanket	The hummocky surface surrounding an impact crater, consisting of excavated material.
convection	The transfer of heat by moving currents of material.
Copernican theory	The heliocentric, or sun-centered, view of the Solar System.
core	The central part of a planet.
cosmic rays	Charged particles (atomic nuclei and electrons) moving in space at close to the speed of light.
crater	A circular depression on a planet.
crust	The relatively thin, outermost layer of a planet. This layer is chemically distinct from the underlying mantle.
degree	1/360 of a circle.
density	The ratio of an object's mass to its volume.
despinning	The slowing down of a body's rotation rate due to the action of tidal forces.
differentiation	A separation or segregation of different kinds of material in different layers in the interior of a planet.
dipole	Any object that is oppositely charged at two points.
discontinuous ejecta	The region beyond the continuous ejecta blanket surrounding a crater, where strings and clusters of secondary craters and rays occur.
eccentricity	The ratio of the distance between the foci of an ellipse and the major (or longest) axis.

ecliptic plane	The plane of the Earth's orbit about the Sun.
ejecta	Material excavated during the formation of an impact crater and deposited around the crater.
electron	A negatively charged subatomic particle that normally moves about the nucleus of an atom.
element	A substance that is made of atoms with the same chemical properties and that cannot be decomposed chemically into simpler substances.
ellipse	A closed curve of oval shape.
elongation	A planet's angular distance from the Sun.
energy	The ability to do work.
erg	A metric unit of energy.
escape velocity	The speed an object must achieve in order to break away gravitationally from another body.
excavation cavity	A crater formed by the ejection of material.
flood basalts	Basalt that forms thick, extensive volcanic plains.
focii of an ellipse	Two points within an ellipse with the property that the sum of the distances from each focus to a point on the curve remains constant.
free surface	An unconfined surface; in an impact event, the surface of a planet or growing excavation crater.
geomagnetic	Referring to the Earth's magnetic field.
graben	A type of trough in which a section of the crust slides downward between two oppositely facing normal faults.

granite	A coarse-grained igneous rock containing quartz and potassium-aluminum silicates.
gravitation	The tendency of matter to attract itself.
gravity assist	The technique of using a planet's gravitational attraction to change the speed and direction of a spacecraft without using fuel.
great circle	Circle on the surface of a sphere that represents the curve of intersection of the sphere with a plane passing through its center.
gyro	A wheel or disk mounted to spin rapidly about an axis that is free to turn in various directions but which resists any motion that would change the axial direction of spin.
Harmonic Law	Kepler's third law of planetary motion: the cubes of semimajor axes of the planetary orbits are proportional to the squares of the sidereal periods of the planet's revolutions about the Sun.
heavy bombardment	The period of time, apparently between about 3.8 to 4.5 billion years ago, when the cratering rate was high throughout the Solar System.
helio	Prefix referring to the Sun.
heliocentric	Centered on the Sun.
hilly and lineated	A type of Mercurian terrain consisting of hills and valleys that disrupt other landforms. Located at the Caloris antipodal point.
hot poles	The subsolar points on Mercury at zero and 180 degrees that face the Sun at perihelion.

impact melt — Target material that was melted by the heat generated by an impact.

inclination (of an orbit) — The angle between the orbital plane of a body and the ecliptic plane.

inferior conjunction — The configuration in which an inferior planet lies exactly between the Earth and the Sun.

infrared radiation — Radiation with a wavelength greater than red light; invisible but felt as heat.

intercrater plains — Level to gently rolling surfaces with a rough texture due to a large number of small superposed craters.

interplanetary medium — The sparse distribution of gas and solid particles in the space between the planets.

ion — An atom that has become electrically charged by the addition or loss of one or more electrons.

ionosphere — The upper region of an atmosphere in which many of the atoms are ionized.

isotope — Any of two or more forms of the same element whose atoms all possess the same number of protons but different numbers of neutrons.

Kepler's Laws — Three laws, discovered by Johannes Kepler, that describe the motions of the planets.

kinetic energy — Energy associated with motion; the kinetic energy of a body is one-half the product of its mass and the square of its velocity.

latitude — Angular distance north or south from the equator of a planet, measured in degrees.

launch window	A range of dates during which a space vehicle can be launched for a specific mission without exceeding the fuel capabilities of that system.
Law of Equal Areas	Kepler's second law: the radius vector from the Sun to any planet sweeps out equal areas in the planet's orbital plane in equal intervals of time.
lithosphere	The outer rigid layer of a planet or satellite.
lobate scarps	Long, sinuous cliffs that cut the Mercurian surface for hundreds of kilometers, characterized by a rounded and lobed appearance.
longitude	Angular distance due east or west from the prime meridian of a planet.
Lunar Orbiter	A series of unmanned orbiting spacecraft launched by the United States to photographically explore the Moon.
magnetic axis	The imaginary line along which a magnetic dipole points.
magnetic field	The region of space near a magnetized body within which magnetic forces can be detected.
magnetosheath	An elongated cavity between the bow shock wave and a planet's magnetic field.
magnetosphere	The region around a planet occupied by its magnetic field.
mantle	The middle layer of a planet, between the crust and the core, and chemically distinct from them.
mare	Latin for "sea"; name applied to many of the dark regions of the Moon.

Mariner space probes	A series of spacecraft launched in the 1960s and 1970s to explore Mercury, Venus, and Mars.
mass	A measure of the total amount of material in a body; defined either by the inertial properties of the body or by its gravitational influence on other bodies.
meteor	The bright streak of light that occurs when a solid particle from space enters a planet's atmosphere and burns by friction.
meteorite	A meteoroid that strikes the surface of a planet.
meteoroid	A meteoritic particle in space before any encounter with a planet.
micrometeorites	Meteorites only a few microns (10^{-4} centimeters) in diameter.
molecule	A combination of two or more atoms bound together; the smallest particle of a chemical compound or substance that exhibits the chemical properties of that substance.
morphology	The shape and structure of a landform.
neutron	A subatomic particle with no charge and with mass approximately equal to that of the proton.
Newton's Laws	The laws of mechanics and gravitation formulated by Isaac Newton.
node	The intersection of the orbit of a body with the ecliptic plane.
normal fault	A fault resulting from tension that pulls the crust apart and causes crustal lengthening.

nucleus (of atoms)	The heavy part of an atom, composed mostly of protons and neutrons, and about which the electrons revolve.
occultation	An eclipse of a star, planet, or spacecraft by a satellite or a planet.
orbit	The path of one object around another body.
perihelion	The closest approach of a planet to the Sun.
perturbation	The deviations from the expected motion of a body caused by the gravitational influence of other bodies.
phase	The fraction of a planet's or satellite's illuminated disk that is visible to an observer.
photometry	The measurement of light intensities.
photon	A discrete unit of electromagnetic energy.
Pioneer spacecraft	A series of spacecraft launched by the United States in the 1970s to Venus and to Jupiter and more distant planets.
planetesimal	Asteroid-sized bodies that in the formation of the Solar System combined with each other to form the protoplanets.
planitia	Latin for plains.
plasma	A hot, ionized gas.
plate tectonics	Motions of a planet's lithosphere, causing fracturing of the surface into plates.
polarization	A condition in which the planes of vibration of the various rays in a light beam are at least partially aligned.

precession	A slow rotation of a planet's axis due to the gravitational effects of a larger body on the planet's equatorial bulge.
prime meridian	The meridian of zero degrees longitude on a planet.
projectile	An object moving at a high speed due to an influence of an external force.
proton	A heavy subatomic particle that carries a positive charge; one of the two principal constituents of the atomic nucleus.
quadrature	A planetary elongation 90 degrees east or west of the Sun.
radioactivity (radioactive decay)	The process by which an element decays into lighter elements.
rarefaction wave	A tensional wave formed by the rebound of rock after passage of a shock wave from an impact event.
ray	Any of a system of bright elongated streaks associated with a crater.
real-time picture taking	Transmission of pictures as fast as they are taken, rather than storage of the images on a tape recorder to be played back at later times.
refraction	The bending of light or radio waves.
relative age	The age of a feature relative to its surroundings (i.e., older or younger than another feature or surface).
relativity	One of two theories developed by Albert Einstein. The special theory treats time and distance as depending upon the motion of the object and the observer. The general theory relates the structure of space to gravitation.

resolution	The finest amount of detail that can be seen on an object's surface.
resonance	See *commensurability.*
rotation	Turning of a body about an axis running through its center.
rupes	Latin for scarps or cliffs.
satellite	A body that revolves around a larger one.
secondary impact craters	Craters formed by the impact of material thrown out during the excavation of a larger crater.
seismic waves	Vibrations traveling through a planet's interior that result from earthquakes or impacts.
shield volcano	A broad, gently sloping volcano built by flows of highly fluid lava.
shock wave	The surface of highly compressed material moving through a medium at an extremely high speed.
sidereal period	The period for one object to complete an orbit around another body.
site-frequency distribution	A graphic illustration depicting the variation in the number of craters at various diameters.
smooth plains	Mercurian plains with a smooth appearance due to a small number of superposed craters.
solar nebula	A cloud of gas and dust from which the Solar System is presumed to have formed.
solar wind	A stream of charged particles, mostly protons and electrons, that escape the Sun's outer atmosphere at high speeds and stream out into the Solar System.

spin-orbit coupling	A relationship between a planet's rotational and orbital periods such that they are whole number multiples of each other.
stereoscopic	A three-dimensional view of a picture, obtained by looking through an optical instrument at two photographs of the same scene taken at slightly different angles.
stratovolcanoes	A steep volcanic cone composed of both lava and pyroclastic material.
stress	Force per unit area.
subsolar	The point on a planet's surface where the Sun lies directly overhead.
superposition	The position of one feature on top of another feature or surface.
surface gravity	The gravitational attraction at the surface of a planet.
synchronous period	A situation in which a planet or satellite always keeps the same face toward the body around which it revolves.
synodic period	The interval between successive similar lineups of a body with the Sun.
T-Tauri stage	An early stage of a star's life, characterized by strong solar winds.
target material	The surface material into which an impact occurs.
tectonics	The study of the large-scale movements and deformation of a planet's crust.
tension	The condition of being pulled or stretched.
telemetry	The transmission of information by radio to a distant station.

terminator	The line of sunrise or sunset on a body.
terrestrial planet	Any of the planets Mercury, Venus, Earth, Mars, and sometimes the Moon.
thermal history	The series of events throughout time caused by the internal temperature of a body.
thrust fault	A fault resulting from compression that pushes one part of the crust over another part and causes crustal shortening.
tide	The deformation of a body caused by the gravitational attraction of a second body.
transection	The cutting of one feature by another.
transit	The passage of a planet (or satellite) across the face of the Sun (or parent planet).
transverse fault	A fault resulting from the horizontal slippage of two pieces of crust past each other.
tsunami	A series of fast waves of seismic origin traveling through the ocean; popularly called "tidal waves."
ultraviolet radiation	Radiation with wavelengths just shorter than violet light and not visible to the human eye.
uncompressed density	The density a planet or satellite would have if the internal material was not compressed to higher densities by pressure.
valles	Latin for valley.
Van Allen radiation belt	A doughnut-shaped region surrounding the Earth where many rapidly moving charged particles are trapped in its magnetic field.

Viking space probes	Two unmanned space probes, each consisting of an orbiter and a lander, launched by the United States in the mid-1970s to explore Mars.
viscosity	A liquid's resistance to flowing.
volatile	A material easily vaporized.
volume	A measure of the total space occupied by a body.
Voyagers	A series of spacecraft that were launched by the United States in 1977 to explore the outer Solar System.
warm poles	The subsolar points on Mercury at 90 and 270 degrees that face the Sun at aphelion.
wrinkle ridges	Ridges that occur on the lava plains of the Moon, Mars, and Mercury, probably due to compression.
zenith	The point directly overhead.

Appendix 2

Orbital and Physical Data for Mercury

Orbital Data

Semimajor axis	0.3871 AU (5.79×10^7 km)
Perihelion distance	0.3075 AU (4.60×10^7 km)
Aphelion distance	0.4667 AU (6.98×10^7 km)
Sidereal period	87.97 days
Synodic period	115.88 days
Orbital eccentricity	0.20563
Inclination of orbit to ecliptic	7.004 degrees
Mean orbital velocity	47.87 km/s
Rotational period	58.646 days

Physical Data

Radius	2,439 km
Surface area	7.475×10^7 km^2
Volume	6.077×10^{10} km^3
Mass	3.302×10^{26} g
Mean density	5.44 g/cm^3
Surface gravity	370 cm/s^2
Escape velocity	4.25 km/s
Surface temperature extremes	90 to 740° K (−183 to 467° C)
Normal albedo (5° phase angle)	0.125
Magnetic dipole moment	4.9 (±0.2) $\times 10^{22}$ gauss cm^3

Appendix 3

Names and Locations of Mercury's Surface Features

Craters	*Quadrangle*	*Latitude (deg)*	*Longitude (deg)*	*Diameter (km)*
Abu Nuwas	H-6	17.5	21	115
Africanus Horton	H-11	−50.5	42	120
Ahmad Baba	H-3	58.5	127	115
Al-Akhtal	H-1	59	97	102
Alencar	H-12	−63.5	104	85
Al-Hamadhani	H-2	39	89.5	170
Al-Jāhiz	H-6	1.5	22	95
Amru Al-Qays	H-8	13	176	50
Andal	H-11	−47	38.5	90
Aristoxenes	H-1	82	11	65
Asvaghosa	H-6	11	21	80
Bach	H-12, H-15	−69	103	225
Balagtas	H-6, H-11	−22	14	100
Balzac	H-8	11	145	65
Bartók	H-12	−29	135	80
Bashō	H-12	−32	170.5	70
Beethoven	H-7, H-12	−20	124	625
Belinskij	H-15	−76	104	70
Bello	H-7	−18.5	120.5	150
Bernini	H-15	−79.5	136	145
Bjornson	H-1	73	110	90
Boccaccio	H-15	−80.5	30	135
Boethius	H-7	−0.5	74	130
Botticelli	H-3	64	110	120
Brahms	H-3	58.5	177	75
Bramante	H-11	−46	62	130
Brontë	H-3	39	126.5	60
Brueghel	H-3	50	108	75
Brunelleschi	H-6	−8.5	22.5	140
Burns	H-3	54	116	45
Byron	H-6	−8	33	100
Callicrates	H-11	−65	32	65
Camões	H-15	−70.5	70	70

Craters	*Quadrangle*	*Latitude (deg)*	*Longitude (deg)*	*Diameter (km)*
Carducci	H-11, H-12	−36	90	75
Cervantes	H-15	−75	122	200
Cézanne	H-7	−8	124	75
Chaikovskij	H-6	8	50.5	160
Chao Meng-Fu	H-15	−87.5	132	150
Chekov	H-11	−35.5	61.5	180
Chiang K'ui	H-7	14.5	103	40
Chŏng Ch'ŏl	H-3	47	116	120
Chopin	H-12, H-15	−64.5	124	100
Chu Ta	H-7	2.5	106	100
Coleridge	H-11	−54.5	66.5	110
Copley	H-11	−37.5	85.5	30
Couperin	H-3	30	152	75
Darío	H-11	−26	10	160
Degas	H-3	37.5	127	45
Delacroix	H-12	−44.5	129.5	135
Derzhavin	H-2	44.5	35.5	145
Despréz	H-1	81	92	40
Dickens	H-15	−73	153	72
Donne	H-6	3	14	90
Dostoevskij	H-12, H-13	−44.5	177	390
Dowland	H-12, H-13	−53	180	80
Dürer	H-3, H-7	22	119.5	190
Dvořák	H-6	−9.5	12.5	80
Echegaray	H-2	43	19	75
Eitoku	H-8, H-12	−21.5	157.5	105
Equiano	H-11	−39	31	80
Fet	H-8	−5	180	24
Flaubert	H-6	−14	72	95
Futabatei	H-7	−15.5	83.5	55
Gainsborough	H-12	−36	183	100
Gauguin	H-1, H-3	66.5	97	75
Ghiberti	H-11	−48	80	100
Giotto	H-6	12.5	56	150
Gluck	H-2	37.5	18.5	85
Goethe	H-1	79.5	44	340
Gogol	H-3	−28	147	87
Goya	H-8	−6.5	152.5	135
Grieg	H-3	51	14	65
Guido d'Arezzo	H-11	−38	19	50
Hals	H-12	−55	115	100
Handel	H-6	4	34	150
Han Kan	H-15	−71.5	145	50
Harunobu	H-7	15.5	141	100
Hauptmann	H-12	−23	180	120

Craters	*Quadrangle*	*Latitude (deg)*	*Longitude (deg)*	*Diameter (km)*
Hawthorne	H-12	−51	116	100
Haydn	H-11	−26.5	71.5	230
Heine	H-3	33	124.5	65
Hesiod	H-11	−58	35.5	90
Hiroshige	H-6	−13	27	140
Hitomaro	H-6	−16	16	105
Holbein	H-2	35.5	29	85
Holberg	H-11, H-15	−66.5	61	66
Homer	H-6	−1	36.5	320
Horace	H-11, H-15	−68.5	52	48
Hugo	H-2	39	47.5	190
Hun Kal	H-6	−0.5	20	1.5
Ibsen	H-6, H-11	−24	36	160
Ictinus	H-15	−79	165	110
Imhotep	H-6	−17.5	37.5	160
Ives	H-12	−32.5	112	20
Janáček	H-3	56	154	47
Jókai	H-1	72.5	136	85
Judah Ha-Levi	H-7	11.5	108	85
Kālidāsā	H-8	−17.5	180	110
Keats	H-12, H-15	−69.5	154	110
Kenkō	H-6, H-11	−21	16.5	90
Khansa	H-11	−58.5	52	100
Kōshō	H-1	60	138	65
Kuan Han-ch'ing	H-2	29	53	155
Kuiper	H-6	−11	31.5	60
Kurosawa	H-11	−52	23	180
Leopardi	H-15	−73	180	69
Lermontov	H-6	15.5	48.5	160
Lessing	H-11	−29	90	100
Liang K'ai	H-13	−39.5	183.5	105
Li Ch'ing-Chao	H-15	−77	73	60
Li Po	H-6	17.5	35	120
Liszt	H-8	−16	168	85
Lu Hsun	H-6	0.5	23.5	95
Lysippus	H-7	1.5	133	150
Ma Chih-Yuan	H-11	−59	77	170
Machaut	H-7	−1.5	83	105
Mahler	H-6	−19	19	100
Mansart	H-1	73.5	120	75
Mansur	H-3	47.5	163	75
March	H-3	31.5	176	55
Mark Twain	H-7	−10.5	138.5	140
Martí	H-15	−75.5	164	63
Martial	H-1	69	178	45

Craters	*Quadrangle*	*Latitude (deg)*	*Longitude (deg)*	*Diameter (km)*
Matisse	H-7, H-12	−23.5	90	210
Melville	H-2, H-6	22	9.5	145
Mena	H-7	0.5	125	20
Mendes Pinto	H-11	−61	19	170
Michelangelo	H-12	−44.5	110	200
Mickiewicz	H-3, H-7	23.5	102.5	115
Milton	H-8, H-12	−25.5	175	175
Mistral	H-6	5	54	100
Mofolo	H-11	−37	29	90
Molière	H-6	16	17.5	140
Monet	H-2	44	9.5	250
Monteverdi	H-2	64	77	130
Mozart	H-8	8	190.5	225
Murasaki	H-6	−12	31	125
Mussorgskij	H-3	33	96.5	115
Myron	H-1	71	79.5	30
Nampeyo	H-11	−39.5	50.5	40
Nervo	H-3, H-4	43	179	50
Neumann	H-11	−36.5	35	100
Nizāmī	H-1	71.5	165	70
Ovid	H-11, H-15	−69.5	23	40
Petrarch	H-11	−30	26.5	160
Phidias	H-8	9	150	155
Philoxenus	H-7	−8	112	95
Pigalle	H-11	−37	10.5	130
Po Chü-I	H-8	−6.5	165.5	60
Po Ya	H-11	−45.5	21	90
Polygnotus	H-6	0	68.5	130
Praxiteles	H-2	27	60	175
Proust	H-6	20	47	140
Puccini	H-11, H-15	−64.5	46	110
Purcell	H-1	81	148	80
Pushkin	H-11, H-15	−65	24	200
Rabelais	H-11	−59.5	62.5	130
Rajnis	H-7	5	96.5	85
Rameau	H-11	−54	38	50
Raphael	H-7	−19.5	76.5	350
Ravel	H-6	−12	38	75
Renoir	H-6	−18	52	220
Repin	H-6	−19	63	95
Riemenschneider	H-12	−52.5	100.5	120
Rilke	H-11	−44.5	13.5	70
Rimbaud	H-15	−63	148	85
Rodin	H-2, H-6	22	18.5	240
Rubens	H-2	59.5	73.5	180

Craters	*Quadrangle*	*Latitude (deg)*	*Longitude (deg)*	*Diameter (km)*
Rublev	H-8	−14.5	157.5	125
Rūdakī	H-6	−3.5	51.5	120
Rude	H-11	−33	80	75
Rūmī	H-12	−24	105	75
Sadī	H-15	−77.5	56	60
Saikaku	H-1	73	177	80
Sarmiento	H-13	−28.5	188.5	115
Sayat-Nova	H-12	−27.5	122.5	125
Scarlatti	H-3	40.5	99.5	135
Schoenberg	H-7	−15.5	136	30
Schubert	H-11	−42	54.5	160
Scopas	H-15	−81	173	95
Sei	H-11, H-12	−63.5	88.5	130
Shakespeare	H-3	48.5	151	350
Shelley	H-12	−47.5	128.5	145
Shevchenko	H-11	−53	47	130
Sholem Aleichem	H-2	51	86.5	190
Sibelius	H-12	−49	145	90
Simonides	H-11	−29	45	95
Sinan	H-6	16	30	140
Snorri	H-7	−8.5	83.5	20
Sophocles	H-8	−6.5	146.5	145
Sor Juana	H-2	49	24	80
Sōseki	H-2	39	38	90
Sōtatsu	H-11	−48	19.5	130
Spitteler	H-11, H-15	−68	62	66
Stravinsky	H-2	50.5	73	170
Strindberg	H-3	54	136	165
Sullivan	H-7	−16	87	135
Sūr Dās	H-12	−46.5	94	100
Surikov	H-12	−37	125	105
Takanobu	H-3	31	108	80
Takayoshi	H-12	−37	164	105
Tansen	H-7	4.5	72	25
Thākur	H-6	−2.5	64	115
Theophanes	H-7	−4	143	50
Thoreau	H-7	6	133	80
Tintoretto	H-11	−47.5	24	60
Titian	H-6	−3	42.5	115
Tolstoj	H-8	−15	165	400
Ts'ai Wen-chi	H-2, H-6	23.5	22.5	120
Ts'ao Chan	H-7	−13	142	110
Tsurayuki	H-11	−62	22.5	80
Tung Yüan	H-1	73.5	55	60
Turgenev	H-1, H-3	66	135	110
Tyagaraja	H-8	4	149	100

Craters	Quadrangle	Latitude (deg)	Longitude (deg)	Diameter (km)
Unkei	H-11	−31	62.5	110
Ustad Isa	H-12	−31.5	166	105
Vālmiki	H-7, H-12	−23.5	141.5	220
Van Dijck	H-1	76.5	165	100
Van Eyck	H-3	43.5	159	235
Van Gogh	H-15	−76	135	95
Velázquez	H-2	37	54	120
Verdi	H-1, H-3	64.5	169	150
Vincente	H-12	−56.5	143	85
Vivaldi	H-7	14.5	86	210
Vlaminck	H-2	28	13	97
Vyāsa	H-2	48.5	80	275
Wagner	H-12, H-15	−67.5	114	135
Wang Meng	H-7	9.5	104	170
Wergeland	H-11	−37	56.5	35
Whitman	H-3	41	111	70
Wren	H-2	24.5	36	215
Yakovlev	H-12	−40.5	163.5	100
Yeats	H-6	9.5	35	90
Yun Sŏn-Do	H-15	−72.5	109	61
Zeami	H-8	−2.5	148	125
Zola	H-3	50.5	178	60

Mountains (Montes)	*Quadrangle*	*Latitude (deg)*	*Longitude (deg)*
Caloris	H-3, H-4, H-8	22–40	180
Plains (Planitiae)			
Borealis	H-1	70	80
Budh	H-8	18	148
Caloris	H-4, H-8	30	195
Odin	H-3, H-8	25	171
Sobkou	H-3	40	130
Suisei	H-1, H-3	62	150
Tir	H-8	3	177
Ridges (Dorsa)			
Antoniadi	H-2, H-6	28	30
Schiaparelli	H-8	24	164
Scarps (Rupes)			
Adventure	H-11	−64	63
Astrolabe	H-11	−42	71

Scarps (Rupes)	*Quadrangle*	*Latitude (deg)*	*Longitude (deg)*
Discovery	H-11	−53	38
Endeavour	H-2	38	31
Fram	H-12	−58	94
Gjöa	H-12	−65	163
Heemskerck	H-3	25	125
Hero	H-12	−57	173
Mirni	H-11	−37	40
Pourquois-Pas	H-12	−58	156
Resolution	H-11	−62	52
Santa María	H-6	6	20
Victoria	H-2	50	32
Vostok	H-11	−38	19
Zarya	H-11	−42	22
Zeehaen	H-3	50	158
Valleys (Valles)			
Arecibo	H-11	−27	29
Goldstone	H-6	−15	32
Haystack	H-6	5	46.5
Simeiz	H-6	−12.5	65

Index